Capitalist Accumulation and Socio-Ecological Resilience

Transpekte:
Transdisziplinäre Perspektiven
der Sozial- und Kulturwissenschaften

Transpects:
Transdisciplinary Perspectives
of the Social Sciences and Humanities

Herausgegeben von Johannes Angermüller,
Anke Bartels, Dietmar Fricke, Raj Kollmorgen,
Jörg Meyer, Dirk Wiemann

Bd./Vol. 9

Edna Yiced Martinez

Capitalist Accumulation and Socio-Ecological Resilience

Black People in Border Areas of Colombia and Ecuador and the Palm Oil Industry

PETER LANG

Bibliographic Information published by the Deutsche Nationalbibliothek
The Deutsche Nationalbibliothek lists this publication in the Deutsche Nationalbibliografie; detailed bibliographic data is available in the internet at http://dnb.d-nb.de.

Registered: Berlin, Freie Univ., Diss., 2016

British Library Cataloguing-in-Publication Data: A catalogue record for this book is available from The British Library, Great Britain
Library of Congress Control Number: 2017043035

Cover Design: © Olaf Gloeckler, Atelier Platen, Friedberg
Cover image: Afro-Black descendant boy in San Lorenzo (Ecuador).
Photo: Edna Yiced Martínez, 2013

D 188
ISSN (Print) 1860-868X
ISBN 978-3-631-73370-7 (Print) · E-ISBN 978-3-631-73853-5 (E-PDF)
E-ISBN 978-3-631-73854-2 (EPUB) · E-ISBN 978-3-631-73855-9 (MOBI)
DOI 10.3726/b12569

© Peter Lang GmbH
Internationaler Verlag der Wissenschaften
Berlin 2018
All rights reserved.

Peter Lang – Berlin · Bern · Bruxelles · New York ·
Oxford · Warszawa · Wien

All parts of this publication are protected by copyright. Any utilisation outside the strict limits of the copyright law, without the permission of the publisher, is forbidden and liable to prosecution. This applies in particular to reproductions, translations, microfilming, and storage and processing in electronic retrieval systems.

This publication has been peer reviewed.

http://www.peterlang.com

Preface and acknowledgments

This book is the product of the academic work submitted in partial fulfillment of the requirements for the degree of: Doctor of Philosophy in Sociology to the Department of Sozial und Politikwissenschaften of the Freie Universität Berlin, in November 2016. I want to thank the many people who have helped and supported me during this process. Thanks to my mother for her love and for everything she did. To my son Juan, he is my best and most loved creation. Thanks to my professor and tutor Sergio Costa for his guidance and trust, and to professor Agustin Laó-Montes who traveled a thousand kilometers to be part of the jury panel for my Viva Voce. As well, thanks to the Rosa Luxemburg Foundation for accepting me in its scholarships program, which was a great experience. I am also grateful for the support of those who helped by correcting and editing the text. Special thanks and gratitude to the Jachnow family who have been my family in Germany. Thanks to Las Kakalaques, to Joachim, to the WG an Kottbusser Damm, and all the rest of my friends for their patience and solidarity.

For Carlos Rua. Brother, friend and teacher, who taught me to recognize in the howl of the core, the tragedy of the periphery. (R.I.P-2015)

Map of the region Tumaco-San Lorenzo and the Colombia–Ecuador border

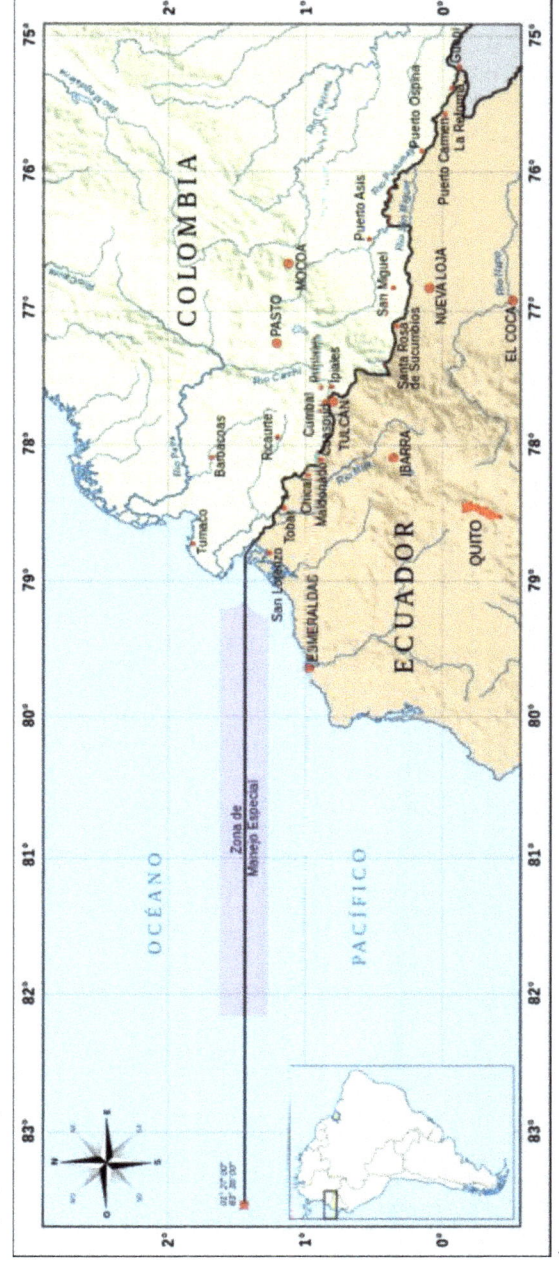

1 From Shadowxfox – Trabajo propio, CC BY-SA 3.0, https://commons.wikimedia.org/w/index.php?curid=22228744

Table of Contents

Introduction 15
    Preliminary Tools 19

I. **Black Peoples' Territories in Colombia and Ecuador: Beyond Capitalist Accumulation and Peripheral Economy** 23
    Redefining the problem: The Colombo-Ecuadorian Pacific coast within global capitalist accumulation 23
    Some reflections on the current issue of "land grabbing" 23
    Global capitalist accumulation 26
    Core-peripheral structure 33
    Primitive accumulation 35
    Disassembling capitalo-centrism 43
    Taking back the economy: Tumaco-San Lorenzo and diversity economic praxis 50

II. **The Long History of the Palm Oil Agribusiness and Development in Tumaco and San Lorenzo** 55
    The Elaeis guineensis 56
    Oil palm expansion: A network of science, money and policy makers 59
    History of the oil palm trade since the fifteenth century 63
        West Africa 63
        South Asia 68
        Central and South America 71
        Colombia and Ecuador 74

History, development and features of the palm oil industry in Tumaco and San Lorenzo 79
    The arrival of the capitalist class, solving the infrastructure leak and the struggle against the "Negros" 81
    Arrival and establishment of palm plantations 84
    The golden era 85
    The bloody era 85
    The bloody era continues 89
Inside palm oil plantations 91
Conclusions 97

III. Imperialism, Unequal Exchange and Palm Oil 99

Imperialism and the palm oil agribusiness 101
Imperialism, nature and the labor force 104
    Nature in the palm oil business 105
    Soil, water, rainfall and sunlight 107
    The labor force: Individual and collective production 109
The magic of capital 114
Conclusions 119

IV. Capital Accumulation and Socio-Ecological Resilience 121

Socio-ecological resilience and capitalist accumulation 124
Production of nature 126
    Geomorphology, climatology and biota 129
    Fauna 132
Production of society 134
    Social metabolism and resilience 137
Cycles of spoliation, social resilience and metabolic rift 139
    First cycle: Slavery, survival and freedom 141
    Second cycle: Manumission, freedom to die and the free market 149

	Third cycle: Upsurge of collective and ethnic identity against exclusion 154
	Metabolic rift 159
	Conclusions 162
V.	**Afterthoughts** 163
	Questioning certainties and trends 165
	Considering the relations between the past and present from a global perspective 167
	New perspectives of classical issues 169
	… and a "classical perspective" on "new issues" 173

Bibliography 179

Index 201

Introduction

> "Ahora estamos en el tiempo de la palma, como antes fue el tiempo del caucho, del banano, después vino el tiempo de las camaroneras, todo eso vino y se fué".
> Eloisa Boca[2]

Every living being has the ability to regenerate when part or all of its structure has aged, has stopped working or has been damaged. Even in adverse situations, cells, organs, organisms and complete ecosystems can, during certain periods of time, partially or completely renew their structure. This book aims to tell the recurrent history of harm, destruction and reconstruction of Black-Afro[3] communities who live on the Pacific coast, on the border between Colombia and Ecuador, specifically in the rural areas of San Lorenzo and Tumaco.

Since the arrival of European conquerors, both the ecosystem and the population have suffered recurrent processes of destruction, but also reconstruction. For three centuries Spaniards represented the political, economic and cultural elite. In the second decade of the nineteenth century, the colonial period ended. Mining and slavery became expensive, and the *Criollos*, the descendants of Spaniards born in America, wanted independence from the Crown. The black population grew, and the *Palenques*, those areas controlled by *Negro* maroons, were an increasing threat to the dominant

---

2   Eloisa is an inhabitant of San Lorenzo, her story appears in the text <u>Territorios, Territorialidad y Desterritorialización. Un ejercicio pedagógico para reflexionar sobre los territorios ancestrales</u> (Garcia, 2010).
3   Inside the activist groups of Latin America there are some debates about both words – black and afro. For some, the correct form should be the word "afro" plus the demonym, for example Afro-Colombian or Afro-Ecuadorian, because it reclaims the African roots and history of the descendants of slaves in America. For others, the complicity of the African power while trading implies a breaking of an African identity, enslaved people were transformed into "blacks", and the concept "black", but not "afro" has been key to the configuration of the racial segregation system. This work does not intend to take part in this debate, and uses the word combination Black-Afro or alternatively Black-African to denote those people descendant of enslaved Africans living in America.

social order (Uribe, 1963; Kalmanovitz, 2008). The scary native groups remained in the region, and the growing black Africans population fought to rebuild their lives and communities in those areas "abandoned" by whites.

But the time to relish freedom proved to be short-lived. After the Spanish colonial power left, new "conquerors" arrived from England, Germany, the United States and Italy, starting a new cycle of exploitation. Today, a new cycle of colonization and exploitation has begun. Even though some conditions and strategies have changed, the motives behind every stage of colonization have remained the same.

This book takes a historical perspective, yet focuses mainly on the newest feature of the process of colonization and spoliation: the palm oil agribusiness. The research analyzes the palm oil agroindustry within the broader history of the dispossession and exploitation suffered by people and nature in this region. It also draws conclusions regarding the conditions that make the repetition of the "primary-primitive" form of capital accumulation possible.

From a *longue durée* perspective, we see that the socio-economic reality in places like San Lorenzo-Tumaco is much more complex than classical, neoclassical and critical economic analysis generally contends. Through the combination of a historical perspective, critical analysis, sociology and intensive ethnographic work, this book portrays the Tumaco-San Lorenzo region as a place shaped by a variety of relations of production, distribution and consumption.

To attempt to study those structures of production, distribution and consumption and their relation with the capitalist system, it was necessary to confront and call into question some rigid postulates and categories in the orthodox liberal and even Marxist critical socio-economic analysis, e.g. concepts such as the "economy", the "worker", "wage-labor", "imperialism" and so on. The text questions the assumption of capitalism as the "core of socio-economic life" as well, and attempts to open up a new spectrum in order to think about and discuss the relations between society and economy. This does not mean omitting or underestimating capitalism's role – quite the contrary! What this research argues is that in the region of Tumaco-San Lorenzo, as probably in many places around the world, different relations of production, circulation and consumption have existed simultaneously, and are to this day, in constant struggle and competition for control over nature and the labor force.

My research shows that ascribing relevance to structures that mainstream economic research has excluded, and social sciences have analyzed as something exotic or residual also help us to understand not only the dynamics of capital in times of a crisis of accumulation, but also its innermost logic.

This text is a journey. The first stop provides us with tools that invite us to dismantle some ideas, which, as I argue, have in some ways paralyzed socio-economic analysis, as well as others that have distorted our understanding of the relationship between capital accumulation, dispossession of nature and agribusinesses. In the first chapter, the text explores the long history of looting and plundering carried out by national and international companies in Tumaco-San Lorenzo. It is shown that the current palm oil agribusiness is just a new face, a repetition of a process performed centuries ago. Then, in order to understand the character and the continuity of this process, the text discusses ideas on the concept of primitive or primary accumulation, as well as the role that non-capitalist modes of production, structures and relations of production play in the accumulation of capital. This first chapter argues that primary or primitive accumulation is the expropriation of a whole circuit of bio-socio-economic relations, which includes the production of nature and human conditions for existence, in order to generate capital accumulation.

The next stop, the second chapter, analyzes the historical development of the palm oil agribusiness, retracing its main phases of expansion and establishment. It exposes why palm oil was key to the rise of the Industrial Revolution in Europe, its role in the configuration of the global order we now face, exposing the network of scientific advances, business practices and policies involved in the palm oil trade from the early nineteenth century in West Africa, later in Asia and finally in Central and South America. Then it focuses on the socio-political framework around the consolidation of Colombia and Ecuador as the biggest producers in Latin America. The research describes and analyzes the expansion and establishment of the palm oil agribusiness in Tumaco and San Lorenzo, showing how it has implied a deep dislocation of the social and environmental conditions that have supported the existence of the Black-Afro population over centuries. The third chapter reveals the continuity of the imperialist scheme that frames agribusiness, focusing on the unequal exchange of this process. It unfolds how the "magic" of capitalist accumulation works, or, how a tonne of palm oil means poverty and destruction for many but also riches and capital accumulation for a select few. The research shows how the ability of nature to produce bio-ecological conditions, as well as of Black-Africans to reproduce by themselves the energy spent during the working day, constitute the fountain of profit and capital accumulation. With these facts as a framework, the text argues that unequal exchange is the basis of capitalist accumulation as a whole, either in its primary-primitive form or throughout expanded production. The latter is ascertained by tracking all of the socio-bio-ecological elements (matter and energy) involved in the production of palm oil as a commodity, analyzing where matter and energy flow, who profits from them and what workers as well as nature itself receive in return.

The fraudulent character of this kind of exchange is expressed by two facts. The first is the unequal distribution and appropriation of surplus value for the production of commodities between the core and the periphery of the system, and the second is the unequal distribution and appropriation of the whole structure of production, which involves nature, human bodies and social skills. As a result of a fraudulent exchange, the wealth produced by the many is appropriated by a select few. The fact that palm oil is being controlled by local capitalists introduces new elements for the analysis of how imperialism works today. I argue that the relation core-periphery is more dynamic and less determined by geographical positions than usually claimed. The way imperialism is portrayed today, i.e., with a fixed notion of where the core and the periphery are located, is contrasted by the fact that core and periphery relations are currently being expanded globally and production relations are being peripheralized. These are issues that require further research.

In this section, how unequal exchange is carried out is exposed by uncovering all of the socio-bio-ecological elements involved in the production of a particular commodity: palm oil. My argument here is that capital has two entities: one material (summary of goods and commodities), which is a product of the combination of the expropriation of nature, appropriation of socio-biological elements that support the labor force, and the direct and indirect exploitation of actual labor. The immaterial entity of capital creates and recreates those conditions that allow a particular group to appropriate, accumulate and profit from the whole of socio-bio-ecological production.

The journey ends with a tour through the territories of Black-Afro communities on the Colombo-Ecuadorian Pacific coast, including the region's history, its forests, its rivers and mangrove trees. Here we explore some elements of the natural history of this place. The research exposes some environmental features that make it propitious for mining and agribusiness. Then the focus turns to the process of settlement carried out by Black-Africans to "dominate" the region in the midst of the most adverse circumstances. The purpose here is to show how, despite recurring damage, people and nature exhibit again and again their power of resilience.

This work does not seek to compare the Black-Afro populations of Colombia with those of Ecuador; rather, it aims to show that they have common histories and face the same challenges. My work is a small contribution to unifying people and territories split up by nationalist political administrations, a process which has been justified by some academics.[4]

Preliminary Tools

Since the arrival of Europeans, Latin America has been integrated into the world-economy in a peripheral or semi-peripheral position.[5] Since colonial times, this part of the world has played a huge role within capitalist accumulation; it has been a satellite-economy and capitalist forces have consequently marked its history.

It is one thing to admit the veracity of this fact and quite another to assume that Latin America has not had other histories and other social and economic structures. From the point of view of classical and critical economic analysis the whole history of the continent is unimportant or not worth knowing. All other socio-economic experiences and structures are neglected or are merely subordinated to the capitalist one. This misunderstanding is clear, for example, in Andre Gunder Frank and other dependency theorists' reflections on the "Indian Problem":

---

4 Randeria (1999) has come up with the concept of "geteilte Geschichte" to refer to the need to overcome the nationalist borders that shape various disciplines, in particular history, and to think from a transnational perspective.
5 See for example Wallerstein, Braudel, Arrighi, Amin, Frank, Cardozo, and other dependency theorists.

> The Indian problem in Latin America is, in essence, a problem of economic structure of national and international capitalist system as a whole ... The problem of the Indians, like that of underdevelopment as a whole, has its roots in the class and metropolis-satellite structure of capitalism and its manifestations are part and parcel of that structure ... [it] lies in his economic relationship to the other members of the society; and this relationship has been, in turn, determined by the metropolis-satellite structure (Frank, 1982: 123).

Frank's argument, although correct, is incomplete. He reduces the whole of the continent's life and reality to the metropolis-satellite or core-semi and peripheral structure. He subordinates the whole history of areas such as the one studied here to capitalism's development. Frank's narrative is problematic because it does not help us understand the complex reality of these places and their peoples. Nor does it help us answer simple but important questions which are vital today in political discussions throughout the whole continent, such as: Why do indigenous communities still exist in Latin America? Why do non-capitalist socio-economic relations still exist? Why does collective land still exist? Why are afro-descendant, native, and farming communities and so on, still autonomously securing material goods (not to mention other goods) outside of wage relations?

Many scholars have attempted to resolve these dilemmas, which are not only theoretical or analytical, but also political, by confining and labeling those structures as traditional, anachronistic or residual. Others have argued that those structures exist because they are functional to the structure of capitalism, but, at some point, they will disappear.[6] But we should ask ourselves what this functionality really means, and how it works. Why, if those economic practices exist prior to the capitalist stage in the economy's historical development, would they exist merely to function within the capitalist one?

Frank's argument expresses a hegemonic narrative in which life and territories built by the Black-African communities in places such as in Tumaco and San Lorenzo are viewed as "mistakes" in what is widely considered the "normal" development of the dominant world economic system. This is very problematic because it reduces whole communities to objects of

---

6   This is the interpretation that can be said to be common to Marx, Luxemburg and the dependency theorists.

exploitation and subjection without projects of their own nor any starting point for their self-emancipation.

Thus, this narrative ignores a huge piece of the past and the present of the continent. It disregards the history and struggle of millions of indigenous, Black-Afro and other communities that, despite having suffered exploitation and dispossession for centuries under capitalism, have still not reproduced capitalist social and economic relationships.[7] We should not forget that these people struggled against the capitalist system before the "class struggle" concept existed and continue to fight today when many sectors of the "revolutionary class" have been neutralized or have simply disappeared.

Leftist authors who study relations between expropriation and exploitation within the capitalist system generally see only one side of this relation: the capitalist side. The analysis of the development of capitalism has helped, without doubt, to create important theoretical and political tools. Nevertheless, at the same time, the system has been endowed with omniscience, omnipresence and omnipotence. Its power has been extended and expanded from the "economic" sphere to every possible dimension. As a result, our capacity to think, imagine, find and create any solutions has been blocked by right and left versions of "TINA (There Is No Alternative) doctrine".

Western economic science, including Marxism, cannot explain other realities outside capitalism because of its historical influence on them. This "capitalo-centrism", as J. K. Gibson-Graham (1996) called it, has occulted a huge part of Latin America's past and present, as well as the struggles of the people. In the end, the capitalo-centrist perspective ignores the power of human beings and neglects other areas and ways of re-creating life (bio-ecological, social, economic).

This work aims to shift the focus to the other side of the relationship, that is, to focus the lens on those excluded from the benefits of capitalo-centrism. It looks at other aspects of relations of exploitation and expropriation to see things that capitalo-centrism could not, attempting to answer simple questions like how people who receive from the capital less than they need to live keep going.

---

7 For example, the growing power of indigenous and other peoples that have revitalized the struggle and, during the last four decades, strengthened their demands for independence and for political, cultural and economic autonomy.

Looking at other aspects of these relations is necessary to find other ways of thinking about and understanding material and social life. It is necessary to expose, as other authors are doing, the weakness and fragility of the capitalist system and to unmask it. However, it is also important to see other revolutionary agents that have been ignored, mainly by Marxist analysis so that we can face, with hope, the problems that affect us today.

Chapter I. Black Peoples' Territories in Colombia and Ecuador: Beyond Capitalist Accumulation and Peripheral Economy

Redefining the problem: The Colombo-Ecuadorian Pacific coast within global capitalist accumulation

This section develops an analytical framework to understand the role played by the region of Tumaco-San Lorenzo within capitalist accumulation. It begins with some reflections on the current "land grabbing" focus, very popular nowadays in studies on land conflicts. I argue that this focus is problematic because it neglects the long-term perspective of land conflicts. Then, I provide a short revision of the presence of the capitalist system in the area of Tumaco-San Lorenzo, seeking to understand the history of capitalist accumulation in Tumaco-San Lorenzo and exposing the role of non-capitalist structures of production.

The guiding questions throughout this first chapter are: i) Why has it been possible for capitalism to continue up to the present in the region of Tumaco-San Lorenzo, using practically the same methods? I.e., a process of exploitation through dispossession that began in the late fifteenth century; ii) What are the conditions for the reproduction of the labor force and of social life in places where wage-labor relations, the main feature of the capitalist mode of production, are not sufficiently developed?; and iii) How has it been possible for people to remain autonomous farmers, fishers and hunters in places where the capitalist system was established many centuries ago?

Some reflections on the current issue of "land grabbing"

Colombia and Ecuador are the fourth and fifth largest producers of palm oil[8] respectively in the world, and rank high among the countries with the most hectares of oil palm cultivated. The oil palm industry in San Lorenzo

---

8   I use oil palm to refer to the plant and palm oil to refer to the product and the business and industry around it.

(Ecuador) started in the late nineties, coinciding with the construction of the Ibarra-San Lorenzo road. According to official data[9], there are 34 thousand hectares of oil palm in Esmeraldas, representing 53% of the country's total, and approximately 14 thousand hectares are planted in San Lorenzo. The oil palm plantation arrived in Tumaco (Colombia) during the sixties, but it developed as an expansive monoculture only in the late eighties.

In this region, between 80% and 90% of the population is Black-African. It is an area with high levels of poverty and exclusion. People are small farmers, fishermen and women, hunters and small-scale mining communities. The large oil palm plantations have transformed the environmental and social landscape on both sides of the border. Many people associate this industry with violence and a loss of kinship between the people of Tumaco and San Lorenzo. Some communal leaders, academics, local authorities and NGO leaders assert that oil palm has led to widespread fear and death, forced displacement, loss of land and autonomy, destruction of forests, poisoning of rivers, paramilitary activities and the destruction of social and organized structures.[10]

But within these plantations there are, in both countries, different visions and discourses. For businessmen, governments, international credit agencies, and even some Black-Afro small farmers, oil palm plantations mean wealth for all: huge incomes for big and small farmers and stable jobs for workers. In addition, they argue that palm oil contributes to the environment, since it helps to reduce global warming, also presenting a viable solution to the peak oil crisis.

The current expansion of oil palm crops in Africa, Asia and Latin America has been an issue of intense debate around the world. Some scholars and activist groups are concerned due to the "new business model" behind oil palm plantations, since it involves private companies and governments

---

9   Asociación Nacional de Cultivadores de Palma Africana Ecuador (see www.ancupa.com).
10 The palm oil agribusiness and its effects on the environment and native communities has been analyzed and denounced in both scholarly research and political activism (some examples of the latter are the works of Escobar, 1996, 2010; Restrepo, 2005).

profiting from the purchase or lease of huge extensions of land, using them for mining and agribusiness.

"Land grabbing" is a catch-all phrase to describe, analyze and denounce the current legal or illegal concentration of land by national and transnational companies, governments, and individuals. Activists and scholars argue that "land grabbing" is mainly a byproduct of neoliberal policies that are transforming land and natural resources into commodities (Borras & Franco, 2012; Oxfam, 2013; UN, 2010). Because of these policies, many people, mainly small farmers and native communities, particularly in the Global South, are being pauperized, dispossessed and displaced.

A summary analysis of the recent expansion of the oil palm agribusiness in the Tumaco-San Lorenzo region – following huge capital investments as well as its negative impacts on local communities and the environment – would point to this place being a clear example of the land grabbing phenomenon. However, after analyzing the economic and political history of the region and the information collected during my fieldwork, the expansion of oil palm plantations and the social dynamic behind this, it in fact merely looks like a new stage within a long history of struggle for the control of land, natural resources and the labor force in the region. Tracking the development of the oil palm agribusinesses – the first large plantations started in the early twentieth century – and some elements used to describe "current land grabbing", i.e., dispossession of native communities, overexploitation of the labor force, and huge investments of capital by multinational companies, one can conclude they have been present in this form for more than a century.

The "land grabbing" theories lack an historical perspective on land conflicts as well as all the conflicts inherent to this confrontation, mainly labor and nature. Land dispossession, violence, transnational capital investment are not new elements within capitalism, they have been present from the beginning of capitalism as the global economic system.

This book, therefore, aims to i) trace, from an historical analysis, the development of the palm oil agribusiness, from the early trade in the mid-eighteenth century in West Africa to the current expansion of the industry in the territories of Black-Afro communities in San Lorenzo and Tumaco, further describing the industry's regional characteristics and analyzing the scientific, monetary and political networks involved in the expansion of this business; ii) to expose the imperialist character of palm oil exploitation;

and iii) to understand the region's socio-economic history and the constant struggles of its people against exploitation and dispossession.

Global capitalist accumulation

In 1916, Rosa Luxemburg (2003: 45–46) wrote:

> Capital needs the means of production and the labor power of the whole globe for untrammeled accumulation; it cannot manage without the natural resources and the labor power of all territories. Seeing that the overwhelming majority of resources and labor power is in fact still in the orbit of pre-capitalist production – this being the historical milieu of accumulation – capital must go all out to obtain ascendancy over these territories and social organizations ... Yet if the countries of those branches of production are predominantly non-capitalist, capital will endeavor to establish domination over these countries and societies. And in fact, primitive conditions allow for a greater drive and for far more ruthless measures than could be tolerated under purely capitalist social conditions.

Luxemburg's paragraph highlights two main points; first, the need of capitalism to repeat the process of primitive accumulation that Karl Marx described as the origin of capital – *"ursprüngliche Akkumulation"*. Second, the pre-capitalist orbit as the sphere in which capital accumulation occurs.

In her text, The accumulation of capital, Luxemburg (2003) argues that, despite the attempts of Karl Marx and Friedrich Engels, as well as numerous other theorists, various questions remain unanswered, including those pertaining to the reproduction of all social capital, meaning the repetition and renewal of capitalist production, the realization of surplus value, and the way in which capital accumulation works. She endeavors to fill this gap by focusing on non-capitalist relations of production, those structures not ruled by wage-labor relations. Her argument is that among both capitalist and non-capitalist modes of production, there is an historical kind of metabolism:

> Capital cannot accumulate without the aid of non-capitalist organizations, nor, on the other hand, can it tolerate their continued existence side by side with itself. Only the continuous and progressive disintegration of non-capitalist organizations makes accumulation of capital possible (Luxemburg, 2003: 397).

The author extends Marx's analytical model of the creation, circulation and accumulation of capital, paying special attention to an issue commonly ignored by classical economic analyses: *the role played by non-capitalist economies for the capitalist one*. Luxemburg's (2003) work reveals that capital accumulation has two faces that have been working together throughout its history: on the one hand we have capitalist accumulation in which surplus value comes from capitalist relations of production, where the capitalist appropriates the worker's surplus labor, and on the other hand there is the accumulation that comes from non-capitalist relations of production, and this is where surplus value is realized, allowing for the accumulation of capital. The author uncovers within her analysis a fundamental feature of the capitalist system: While capitalism is a mode of production, it is also, above all, a system of accumulation which integrates many structures, modes and relations of production.

Unfortunately, Luxemburg's postulations were long forgotten, and for decades scholars and political movements focused on capitalism strictly as a mode of production; at the same time, Marx's analytical model became a kind of theology. Thanks to the expansion and specialization experienced in European societies in the nineteenth century, Marxists, and even non-Marxists, believed to have found the true identity of capitalism in its industrial stage. This represents an analytical mistake, as Giovanni Arrighi (2000: 5) points out:

> In certain periods, even long periods, *capitalism did seek to "specialize"*, as in the nineteenth century, when it "moved so spectacularly into the new world of industry". This specialization has led historians in general to regard industry as the final flowering which gave capitalism its "true" identity. But it is a short-term view.

The idea that capitalism and industrialization were synonymous generated other conceptual errors, for example the notion that expanded reproduction and surplus value are the same, or that wage labor, the bourgeoisie and the proletariat were the only central categories for understanding production relations within the capitalist system. Therefore, other social and economic experiences remained ignored, and the analysis of capitalism was reduced to its industrial phase. Consequently, the long history of capitalism vanished, as well as its role in the early configuration of what today we call Latin America, generating distortions and the pointless discussion of which phase of production relations Latin America should be framed in: the feudal one, a capitalist (industrial) one, or something in between.[11]

---

11 Carlos Mariategui is one author who early in the twentieth century had already developed a global understanding of the capitalist system. Mariategui

In the sixties and seventies, neo-Marxist scholars such as Andre Gunder Frank, Fernand Braudel, Immanuel Wallerstein, and Giovanni Arrighi recovered Luxemburg's analysis. For these scholars, capitalism is an approximately five century old "world-economy", a *world-system*, featuring two main characteristics: i) the imperative to produce and to sell on the market in order to obtain the maximum profit (Wallerstein, 1974: 13); and ii) the ability of capital to increase continuously (Braudel, 2006; Arrighi, 2000). Within this system, surplus values are produced by different labor regimes such as slavery, servitude and wage labor. The economic system is organized in three interdependent levels: core, semi-periphery and periphery, and supported by many different political and cultural structures. Within the system, there is a functional and geographical division of labor and power that ensures the control of profit by the core states (Frank, 1967; Arrighi, 1999; Braudel, 2006; Wallerstein, 1976).

Since the Europeans' arrival in America, the Pacific coast of Colombia and Ecuador has suffered the effects of this early expansion of the capitalist system. Tumaco-San Lorenzo and the rest of Latin America have served as a kind of pantry for the capitalist core from colonial times to the present. Despite the fact that it was the last area conquered by the Spaniards, most of the colonial gold production during the reign of the Spanish Empire originated from the Pacific region (Leal & Restrepo, 2003). Precious minerals were first extracted by the enslaved indigenous populations, then by slaves kidnapped from Africa. Gold produced in this area made Genoa's and Holland's bankers richer. The gold exploited there bought fine silks and spices from the Far East, as well as ships, sugar, cotton and more slave labor. Colonial products and the terms on which they were produced meant a level of capital accumulation never seen before, accumulation which boosted

---

understood the role played by Latin America within the world-system and the articulation of different regimes of exploitation and control over the labor force. The author's lucidity on these issues can be seen in contributions such as Siete Ensayos sobre la Realidad Peruana (first published in 1928) and Ideologia y Política (first published in 1929). Unfortunately, and maybe due to the eurocentrism that even today dominates intellectual life, Mariátegui's work has long been ignored in the Marxist tradition. Nevertheless, in the last decade there has been a growing interest in recovering Mariategui's thought on the scholarly as well as the political stage.

the Industrial Revolution and the consolidation of British economic and political power during the eighteenth and nineteenth centuries (Williams, 1947; Tirado, 2000).

The colonial wars, the high prices of gold production and the formal prohibition of slavery meant the end of the colonial exploitation cycle in San Lorenzo and Tumaco. Many white Spaniards, former owners of mines, returned to Europe or migrated to the Andean areas (West, 1956). At the same time, Black-African people became the majority of the population by obtaining more territorial control (Taussig, 1979). Former enslaved and self-liberated people organized their lives by developing a balanced economic system with the natural resources available to them: fishing in the low areas, planting in the middle ones, and mining in the high ones (Leal & Restrepo, 2003).

Capital accumulation through colonial exploitation ended, but with the "independence" of the colonies it continued through the world market now dominated principally by England. After independence from the Spanish Crown, British, German and US American capital arrived in Latin America. The new agents of capital, in alliance with the few whites in the region, arose to form a new type of elite that arrived in San Lorenzo and Tumaco in search of "new products" for the international market.

This new wave of colonization and exploitation began with the German-British company Ecuador Land Company Limited (ELCL). This company arrived in 1860 to collect the debts contracted by Ecuador during the wars for independence (Fischer, 2000; Kirchberger, 2013). The Ecuadorian government provided ELCL 1,000,000 *cuadras* (640 million square meters) in San Lorenzo in order to discharge its debts with the British government. The company occupied the land with British and other European settlers with the purpose of transforming the area into a modern port that would connect Ecuador with other ports in America, simplifying the exploitation and trade of products such as timber, tobacco, corn, quinine, cotton, cocoa, coffee, Indian rubber, and cochineal, as well as the exploitation of gold, silver, quicksilver, copper, and emerald mines (Arosemena, 1991; Kirchberger, 2013).[12]

---

12 Though the British-German colonization project was not officially extended to Tumaco, both groups had an important presence in this village, and in tandem

By 1870, the euphoria of that phase of colonization ended. Internal problems within the company, difficulties in developing the project, a weak infrastructure, political instability and poor institutional practices for guaranteeing and protecting intensive capital investment forced the ELCL to liquidate its assets (Fisher, 2000; Arosemena, 1991). But these failures did not signify the end of the arrival of men looking to make a fortune. In 1874, the Grindale Company – founded on the remains of ELCL – arrived in San Lorenzo to manage the exploitation of "vegetable ivory" (*tagua*) from the palm trees and its commercialization. Despite many attempts, the project failed for almost the same reasons as ELCL. The company was finally closed in 1896.

The project of constructing a large commercial port was not carried out; however, significant amounts of money were accumulated, especially with the trade of natural resources such as tobacco, vegetable ivory, and rubber.[13]

The owners of the capital had access to a very cheap labor force, and had monetary, political and armed control over the region. At the same time, they did not have to invest much in building infrastructure.

Black-Afro populations collected the seeds and extracted the rubber. They earned some money buying goods that they could not produce themselves, such as clothes, salt, and tools. However, they also had a strong economic structure based on collective and autonomous work (fishing, hunting, farming) that guaranteed them the majority of goods necessary to survive (West, 2005: Leal, 2008). Consequently, neither traders nor supervisors had to pay wages or make large monetary investments in order to access the vegetable ivory plantation; they only had to create a warehouse and buy small boats to transport the goods (Leal, 2005).

For three decades, British and German companies, and later on companies from Italy and the United States, together with the small white elite minority constituted a kind of parallel state with their own currency and armed forces that guaranteed their control of the region and its trade. Traders took control of the land while monopolizing the trade of the goods,

---

with the local elites, controlled the most important market sectors in Tumaco and San Lorenzo, both the legal and illegal ones.

13 Rubber in San Lorenzo and Tumaco met 75% of the global demand at the end of the nineteenth century.

developing a farm system based on debt peonage; the collectors received credits to acquire goods in the companies' stores. The debts quickly increased and forced the people to work more for the companies until they could completely pay off their debts, which was almost impossible. Those who could not pay were enslaved (Fisher, 2000; Christopher, 1985).

The end of the rubber and vegetable ivory boom meant the arrival of new conquerors, businessmen who reproduced or created new ways of controlling the land and labor force, framed by new relationships with recently formed states. From 1880 until the first half of the twentieth century, there was a new cycle of gold exploitation, this time under the leadership of the United States (Jurado, 1990; Corsetti, 1990). Some years afterwards the extraction of mangrove, tanino, and other native trees began, which were shipped mainly by US companies (Whitten, 1965). Soon after, the export of the native and legendary naidi palms [Açai] to Europe, mainly to France, and the rise of the cocoa plantation along the entire Ecuadorian coast, meant the establishment of a crop that is still one of Ecuador's most important exports (Guerrero, 1983). We should not forget the role played by transport and infrastructure, because construction of ports, railways and roads, headed by the French, English and Italians meant the destruction of nature, loss of land and autonomy for the communities, exploitation of the native labor force, but also wealth for a few, mainly foreigners.[14]

After the Second World War, the "Banana Boys" arrived from the United States, and Ecuador became the largest banana exporter in the world. During the middle of the last century, lumber exportation, another important business that began in colonial times, became one of the most important

---

14 In 1920 a contract was signed for the construction of the railway to Nariño (Pasto-Tumaco) between the governments of J. Buccheli and E.D. Wright in Panama. In 1924, the destruction of the forest began with 30 workers. Due to differences between the regional and national governments concerning the cost-benefit relationship, the project was abandoned. San Lorenzo received credits from the United States to build San Lorenzo's Port, then from France to build railways contracted by the enterprise CIAVE and the Metropolitaines et Coloniales. Between the 1930s and 1960s, there were many conflicts between the Junta Autónoma del Ferrocarril Quito-San Lorenzo an organization encouraged by the train administration, and Black-African communities due to the fact that the Junta had sold ancestral and collective lands to a French company (Whitten, 1965).

elements within its market economy. At the same time, a developing fishing industry, mainly centered on tuna, started to grow rapidly (Florez, 1983). In the 1980s, spurred by public and private investment, intensive shrimp production started, ushering in along with cocoa plantations and the palm oil agribusiness a new stage of capital accumulation in the region.

As can be seen, from colonial times, the Colombian and Ecuadorian Pacific coast has been a place of capital accumulation and has been integrated into the international market. Every cycle concerning the exploitation of nature and labor has had a counterpart of resistance and reconstruction. The Tumaco-San Lorenzo region has been the site of a constant search for freedom – a place where natives and Africans fought together, in many cases, against colonization and slavery. They organized through strong kinship bonds and developed many strategies to confront the interests of the owners of capital, the effects of every exploitation boom, and to recover from every capitalist cycle of accumulation.

Tumaco and San Lorenzo became an area of constant confrontation between different relations of production, circulation and consumption. Here, different visions of the world and distinct forms of the relationship between human beings and nature have been in constant confrontation. Many people arrived looking for riches, taking whatever they could to make money, building basic infrastructure to achieve this goal and abandoning the region when the resource was depleted or was no longer profitable. But others, the natives, who have been there since ancestral times, and the Black-Africans, who arrived as slaves centuries ago, have recreated their life in these areas based on the collective and equitable distribution of land and nature.

Tumaco-San Lorenzo has never been an isolated area within the global accumulation dynamic. This region has played an active role, not only in the economic progress of the countries of Colombia and Ecuador, but also in meeting the demand of international markets. At the same time, the political and economic issues that have shaped the world over the last few centuries have also molded it.[15]

---

15 It is worth mentioning that during the 1960s there was an active presence of communist and socialist activists in San Lorenzo, and in the eighties Tumaco also saw mobilizations by people, many of whom were part of Marxist organizations, demanding basic rights such as clean water. Through interviews, the presence

The above paragraphs briefly describe the socio-economic history of the region and some of the relations, structures and systems of production, distribution and consumption that have operated in it. We see that groups of people with different and even opposing interests have lived together and have profited, at different levels, from the richness of this area. How has it been possible that dissimilar interests could be articulated to generate, up to the present time, profit for capitalists on the one hand, and living conditions for Black-Africans outside of the wage-labor structure on the other? How has capitalism as a mode of production and system of accumulation been working in this zone? What is the role that non-capitalist forms of production play in the development of capitalism as a system of accumulation?

In the next section, I use some concepts developed by Marxist authors such as *core-peripheral* and *primitive accumulation*, exploring their possibilities and limitations in understanding the socio-economic structures in the region of Tumaco-San Lorenzo.

Core-peripheral structure

Expanded reproduction, in which capitalism appropriates surplus value produced by the wage labor scheme within a context in which the workers only have their labor power to sell, has not been the basis of capital accumulation in Latin America. In this part of the world, capital accumulation has been based principally on the dispossession of land and the exploitation of natural resources, as well as an unwaged labor force.

Here, the land is vast and the labor force has either not been free or has been very cheap. The accumulation of capital in Latin America has not transformed non-capitalist modes of production into capitalist ones, and, with a few exceptions, there was not a large industrialization process here. If we understand this merely as the exploitation of wage labor, it has, instead, created peripheral economies specialized in supplying cheap labor and raw materials under many labor regimes to the system's core.

As Samir Amin (1974) has noted, in Latin America, surplus value is created through many different non-capitalist modes of production, with

---

of socialist, communist and even fascist ideas in the region became evident (for more on these issues, see Estupiñan 1991, 1994, 2003; Whitten, 1965).

non-wage-labor institutions such as slavery, the *encomienda* system, and peonage generating higher rates of accumulation for the core of the system than elsewhere. This relationship between the core and periphery was named "unequal exchange", a transfer of products with unequal value and unequal prices of production that, consequently, provide an advantage to the core (Amin, 1974; Wallerstein, 2013). Through this unequal exchange, the "developing" countries supply capital to the advanced ones, not vice versa (Amin, 1974), as is generally claimed.

Capitalism is a market-oriented system, which in Latin America has operated through mainly extractivist activities, including mining, oil and gas drilling and agribusiness. The main goal of these activities is to produce raw materials and food for the external market at a lower cost, generating large concentrations and monopolies of land, money and technology (Amin, 1974; Frank, 1982, Leal & Restrepo, 2003).

In the area this research is centered on, the market-oriented system with its respective capitalist accumulation has been performed through the violent appropriation of natural resources (mountains, forests, water, plants and seeds), but also the appropriation of the labor force, which exists outside wage-labor relations, though there has been resistance from local communities. In this way, the market-oriented system profits not only from nature and people, but also from the ability of communities to keep some kind of balance between man and nature, conserving particular environmental conditions.[16]

---

16 In Latin America, rural populations have fallen drastically in the last decades. In Colombia, this process has been accelerated during the last 20 years because of paramilitary actions in alliance with multinational corporations (approx. two and half million of the displaced population are farmers). Nevertheless, in Colombia and Ecuador the farming population still represents 40% and 30%, respectively, of the whole population. It is also important to point out that, parallel to the urbanization process, the so-called "ethnic groups", such as natives, indigenous and afro-descendants have begun demanding particular rights for the land as well as respect for their socio-economic structures. Additionally, these groups have strongly opposed both extractivism and agribusiness. The opposition to the latter has resulted in significant persecution and violence against these communities. This situation is described in works such as <u>La tierra contra la muerte. Conflictos territoriales en los pueblos indígenas de Colombia</u> (CECOIN, 2008).

In Tumaco-San Lorenzo the most important elements of extractive activities, the environment and the labor force, are produced outside of capitalist relations of production. This does not mean that the owners of capital do not have to invest money in purchasing tools, machines and so on, because they still do so. The point is that, within mining, oil and gas drilling and agribusiness, the relationship between what Marxists call *variable* and *constant capital*[17] has other implications, due to the fact that the owners of capital do not have to pay any money beforehand to obtain natural resources and an available labor force. Both these elements existed previous to the arrival of the capitalist, and, most importantly, *the success of the capitalist enterprise in this part of the world depends on the capacity these elements have to exist free of charge to the capitalist system*. Consequently, in Latin America the capitalist system has waged, for over five centuries, a continuous and constant spoliation, which in Marxist literature is called primitive accumulation.

Primitive accumulation

Primitive accumulation has received renewed scholarly attention, and my thesis is to some extent a product of this renewed interest. Nevertheless, the deeper my bibliographic work went, the clearer the confusion existing on this topic became. In this subchapter, I briefly explore some recent ideas and discussions on this topic and address some of the ideas of authors that I have found to be relevant to the present discussion on the role played by the continuous presence of primitive accumulation within capitalism. Then, I argue that a "forgotten" conceptualization of "primitive", or rather "primary" accumulation developed by Andre Gunder Frank (1978) provides us with good tools to analyze the current global situation, in particular for the analysis of the case-study.

To begin with, it must be stressed that English and Spanish translations of Marx's concept of *ursprüngliche Akkumulation* have increased

---

17 Constant capital is the value of goods and materials acquired by the capitalist required to produce a commodity, while variable capital is the wages paid for the production of a commodity.

the confusion concerning the concept[18], and although discussing linguistic issues would mean straying from the topic, it is relevant to know where the debate starts.[19] Secondly, it is important to note that the debates around this issue are not new, for almost one and a half centuries there have been many interpretations and analyses of what "primitive accumulation" is or was.

In the English version of Capital, "primitive accumulation" is described by Marx as "the historical process of divorcing the producer from the means of production. It appears as primitive, because it forms the prehistoric stage of capital and of the mode of production corresponding with it" (2004: 506). Then, Marx exposes the historical movement, the political and economic conditions that transform producers into wage-laborers, and the violence inherent to this process.

All the later discussions on primitive accumulation have been in some way been based on Marx's analysis. Some authors argue that primitive accumulation was an historical event in Europe that marked the transition from the feudalist mode of production to capitalism. Included here are the majority of the so-called "Orthodox Marxists". For example, Lenin understood this process as the foundational period of capitalism, not only in Europe, but everywhere where capitalism was implanted. For Lenin primitive accumulation is a process through which non-capitalist societies transform into capitalist ones, and once this transition has been made it does not play any further role in the capitalistic structure: primitive accumulation implies the transition from a feudalistic mode of production to a capitalist one.[20]

Others authors argue that "primitive accumulation" has been a constant process within the capitalist system. They contend that the current situation

---

18 With regard to this debate Perlmann (2003) wrote "Although primitive accumulation was an essential concern of classical political economy, the study of this concept began in confusion and has since settled into an unfortunate obscurity. The seemingly Marxian expression, 'primitive accumulation', originally began with Adam Smith's assertion that 'the accumulation of stock must, in the nature of things, be previous to the division of labor' (Smith, 1776, II.3: 277). Marx translated Smith's word, 'previous', as *ursprünglich*' (Marx 1887: 33, 741), which Marx's translators, in turn, rendered as 'primitive'."
19 See Enclosures, The Mirror Image of Alternatives, The Commoner 2, September 2001.
20 See Lenin V. (1975). El Desarrollo del Capitalismo en Rusia.

worldwide – the displacement of native communities, millions of people having their land expropriated, and other means of production, common goods and social rights now being used by vested interests in the pursuit of profit – seems to confirm the idea that we are living through a similar process to Europe during the fifteenth century, a process that, for Marx, was the birth of capital, i.e., primitive accumulation.

One of the authors active in this discussion is Massimo De Angelis. De Angelis (2001) suggests that primitive accumulation was not only a relevant factor in capitalism's initial phase, but is also present in "mature" capitalist systems and, given the fact that he recognizes the centrality of social relations of production and class struggle, assumes a "continuous" character. The same approach is taken by Klaus Dörre, who, although he does not use the term primitive accumulation, uses one that is closely related: *Landnahme*. Dörre argues that Western societies are suffering a second wave of dispossession and exploitation, similar to the one that occurred during the seventeenth century, only that it is now being spearheaded by finance capital. This new wave of exploitation, says Dörre (2010), has two main principles: the dispossession of rights conquered by the working class throughout the last centuries, and the weakening of workers' consciousness.

If the continuance of capital accumulation only exists where and when capitalist relations of production and the process of class formations – proletarians and capitalists – have already been completed, how do these authors explain the exploitation and dispossession suffered by societies in the periphery that for five centuries have been part of the global process of capital accumulation, generating high rates of surplus-value appropriation under other relations of production such as slavery, servitude, and so on?

David Harvey is an author who has reignited interest in the continuity of primitive accumulation and its role inside capitalism as a whole. Harvey argues that primitive accumulation has shaped the history and development of capitalism. He argues that, since the seventies, because of the global crisis of capital overaccumulation, which is the relationship between the increase of capital-money and the contraction of industrial production as the dominant area for investment and accumulation, capitalism lies idle with no profitable outlets in sight. The crisis forces capital to search and find alternatives for keeping and increasing the rates of profit.

The way in which capital deals with its current overaccumulation crisis is no longer through industrial production but rather primitive accumulation, or as Harvey (2005: 146) calls it, *accumulation by dispossession*:

> Primitive accumulation, in short, entailed appropriation and co-optation of pre-existing social and cultural achievements as well as confrontation and suppression. ... The resulted is often to leave a trace of pre-capitalist social relations in working class formation and to create distinctive geographical, historical and anthropological differentiations in how a working class is defined. ... What accumulation by dispossession does is to release a set of assets (including labor power) at very low cost (and in some instances zero cost). Overaccumulated capital can seize hold of such assets and immediately turn them to profitable use. In the case of "primitive accumulation" as Marx described it, this entailed taking land, say, enclosing it, and expelling a resident population to create a landless proletariat, and then releasing the land into the privatized mainstream of capital accumulation.

Harvey argues primitive accumulation or accumulation by dispossession should be understood as a result of dialectical "inside-outside" relations within the capitalist system. Due to capital constantly requiring something "outside of itself" to accumulate when "capital accumulation through expanded reproduction" is not enough during times of crisis, capitalism makes use, often by force, of "pre-existing" outside structures such as non-capitalist social formations or sectors that have not been proletarianized.

Harvey's proposal has been welcomed with great enthusiasm in academic and activist settings. But in trying to use those ideas in my own research, I encountered some problems. Harvey argues that accumulation by dispossession is a fundamental part of the entire capitalist accumulation process. But, in the end, for the author it exists because of a crisis of overaccumulation of capital. In Harvey's analysis primitive accumulation or accumulation by dispossession is a result of the crisis and not an intrinsic part of capitalism.

Harvey uses the term accumulation by dispossession to explain *every current case* of expropriation and exploitation around the world. Yet, by generalizing, he hides the particularities of each process; the loss of workers' rights in Germany, the United States or England is not the same as the theft of natives' lands in America, Asia and Africa. These processes have different histories and consequences. Another problem is that Harvey builds his argument around the analysis of a very short period of time, i.e., the domination of the United States after the Second World War, and neoliberalism. This makes it difficult to understand how primitive accumulation has functioned throughout modern history, particularly in the periphery of capitalism.

I also find it very problematic that Harvey thoughtlessly interchanges the concept of primitive accumulation with accumulation by dispossession. Despite the debates about the exact meaning of the word "primitive" in Marx's work, it is clear to me that "primitive", "*ursprünglich*", and even "primary" are words used to refer to an *origin* or *beginning* of some kind within Marxist thinking, a *transition* and *transformation;* something can be defined as "primitive" or "*ursprünglich*" only in relation to something that it is not. Harvey's concept seems to cover over this issue.

An author who connects core and peripheral societies and primitive accumulation with an historical analysis is Silvia Federici. From a feminist perspective, she focuses on the transitional stage from non-capitalism to capitalism in Europe, America and Africa. Federici (2010) argues that primitive accumulation began prior to the fifteenth century as part of a counterrevolutionary process that meant, on the one hand, the alienation of people's means of living to increase capital accumulation and concentration of labor and money, and on the other, the creation and exploitation of *differences* within the labor force. By the creation of gender and also racial, hierarchies, it was possible to shape a dominant European, masculine elite, as well as a modern proletariat. In this way, Federici reveals how primitive accumulation meant the alienation of labor power, the means of production and the production process, as well as the alienation of humans as social beings and the loss of power and autonomy for women in Europe and for non-European societies around the world.

Federici's work brings interesting elements into the debate, and proposes many ideas to understand how primitive accumulation has operated historically, especially by controlling women's decisions and bodies, and the violence this process involved. But once outside the historical field, Federici's work about current exploitation and appropriation, for example her work on African societies, is more descriptive than analytical, and does not use the richness of her theory about primitive accumulation to explain the current situation of expropriation.[21] How is the continuity of

---

21 See for example her work on the new enclosures in Africa (Federici, 2001).

primitive accumulation possible today, in contexts where the alienation of the human being, the domination of women's bodies through the "witch-hunt," and colonization and spoliation were accomplished during the earliest stage of capitalistic expansion and the consolidation of the capitalist mode of production? Was primitive accumulation as described by Federici an unfinished process and are we now experiencing its continuation? Or, on the contrary, is it a new kind of primitive accumulation that we suffer nowadays, and if this is true, what would its features be, and how would it be linked with the former?

Liza Grandia (2007) has brought new views and concrete examples of how primitive accumulation has been a permanent process in societies with non-capitalist relations of production. Using the example of indigenous communities in Guatemala, who for almost five centuries have constantly lived inside "enclosures" and experienced expropriation and many stages of dispossession, Grandia argues that primitive accumulation has four key characteristics: i) separation from subsistence and new subjectivities; ii) the rule of law and historical amnesia; iii) the collateral damage of debt and never-ending consumerism; and iv) the plunder of resources, disguised as civilized economic development and the banner of "progress". In short, primitive accumulation here means the dispossession of the means of production and the labor force and also the configuration by force of a cultural hegemony.

Grandia's work opens new windows for understanding the permanent processes of dispossession, exploitation and enclosures suffered in places inhabited by communities not subordinated to the wage-labor relation, such as the native communities in Central America. But many points in her argument lead to confusion. She argues that there is not a single form of capitalism, but a plurality of capitalist structures, e.g. "the new 'gangster', capitalism" of Russia, the "crony capitalism" of white South African elites, the "military-corporate capitalism" of the United States, the "dictatorial capitalism" of China, the speculative "dot.com capitalism" of Silicon Valley, in general, the "socialized capitalism" of Europe, (and) the "colonial-cattle capitalism" of Guatemala's "frontier" (Grandia, 2007: 78)

The author neglects capitalism as a world-system and therefore gradually loses analytical sharpness by mixing different structures of surplus-value. Another problematic point with Grandia's work is related to the four characteristics of primitive accumulation she mentions: Should they be assumed as constant features of every kind of capitalism she describes? What is the connection between these diverse capitalist forms she describes and the situation of the native communities in Guatemala and Belize? Is primitive accumulation an intrinsic feature of capitalist expansion or a particular experience lived by native communities in Central America? In addition, the perspective of Grandia's work ends up as an unbalanced one, despite the fact that she has been working closely with non-capitalist communities and has herself witnessed how primitive accumulation is a perpetually contested and challenged process, she does not achieve her goal of dismantling the idea of capitalism as the core of socio-economic cultural structures. Her work fails to provide a theoretical tool to understand the processes of contestation as well as the large-scale confrontation of the capitalist system carried out by native communities throughout the centuries. In short, she does not answer a key question: why and how indigenous communities have resisted and survived every stage of primitive accumulation.

Why it has been possible for capitalism to repeat primitive accumulation again and again around the world for six or more centuries is a question that Grandia, Federicci, Harvey and other authors engaging with the analysis of primitive accumulation do not contemplate; it is a question that they simply omit. Harvey does not seem to care, for example, where the "pre-existing outside", as he calls it, necessary for current capitalist accumulation by dispossession, is coming from. Grandia, for her part, neglects to explain why after more than five centuries of colonization, indigenous communities still exist in Guatemala? Nor does Federici explain, as she notes, that despite the long history of dispossession "to this day at least 60 percent of the African population lives by subsistence farming" (2001: 34).

Andre Gunder Frank's work could help us fill in some gaps left by those authors' concepts of primitive accumulation. To Frank, capitalism as an accumulation system profits in peripheral countries and societies through the exploitation of natural resources and the labor force for which owners of capital have not paid a dime. Frank calls this type of accumulation "primary" to distinguish it from "primitive", in this way referring to capital accumulation based on a non-capitalist mode of production, which means those relations of production not regulated by wage labor.

The production of use value through a non-capitalist relation of production, argues Frank, continues today and is still directly converted into value for the process of capitalist accumulation in regard to three functions: i) the sustaining and, in times of need, provision of a potential reserve army of labor and pool of labor power; ii) the contributions and the sustenance and reproduction of wage labor that is productive, for which capital pays a wage that is too low for the wage laborer's survival and reproductions; and iii) the use of non-capitalist relations of production to produce value that enters into the global capitalist process of capital accumulation (Frank, 1978: 201).

Frank has some doubts as to whether exploitation and non-capitalist means of production for capital accumulation today are as important currently as they were in the past. The large wave of academic and political discussions and debates on "primitive" or rather "primary" accumulation shows how important the issue still is, as well as the centrality of these relations in the process of capital accumulation over the last four decades of neoliberalism.

Academics writing on permanent and current "primitive accumulation" are really talking about "primary accumulation" as Frank conceptualized it. But, despite the general understanding of the importance of non-capitalist, pre-capitalist, and non-proletariat structures inside this process, academics have not given them enough attention within their analyses.

If capitalism appropriates values produced by non-proletarianized labor under non-capitalist means of production through primary accumulation, it means that around the world there exists a huge amount of non-proletarianized labor as well as a large amount of goods that are produced under other relations of production. This also means that these relations of production are very important and significant, despite the fact that an advanced capitalist mode of production has been in existence for more than two centuries. Nevertheless, we know little about the nature of these relations of production and where they come from. There also exists an information gap concerning how it has been possible that, after centuries of hegemonic global expansion, the capitalist system today has not devoured non-capitalist, non-proletarianized areas, societies and communities. Correspondingly, the role these structures have been playing within the market-oriented system remains a gray area.

In the next section, based on the ideas of some of the newest and heterodox Marxist theorists, I attempt to argue that: i) capitalist accumulation, either as expanded production or primary accumulation, is based, in essence, on the exploitation of non-capitalist relations of production; ii) capitalism damages all types of non-profit, non-capitalist, non-proletarianized relations without destroying them at all because it needs them, and iii) those non-capitalist, non-proletarianized relations of production are structures with permanent resilience and regenerative skills.

Disassembling capitalo-centrism

Some kinds of parasites exhibit a high degree of specialization. Some of them even have the ability to take the place of their host, making it hard to recognize the parasitic relation and thus also to remove the parasite from the host. Capitalism as a mode of production and as a system of accumulation has a parasitical nature, which for centuries has been profiting from a universe of structures of production which are not ruled by a profit-oriented relation.[22]

For centuries, the defenders of the capitalist system have exposed it as a self-reproducing structure, which despite some moments of crisis, has, up to the present day, been able to reproduce and re-structure itself. From this perspective, capitalism is a system, which despite the large list of problems around the world, has supposedly created and sustained the most developed type of social order known to man. This discourse is an ideology that has been incorporated into the academic sphere, even within its critical sectors. The purpose of this section is to dismantle the idea of capitalism as a triumphal and omnipotent system. In other words, I want to argue, as Karl Polanyi did but with another focus, that the basis of the existence of capitalism has been the recurrent and literal dispossession and dislocation of the environmental and social order.[23]

To dismantle the idea of capitalism as omnipotent and triumphant, I use some elements of Michel Foucault's genealogical method (Foucault, 1977), striping capitalism of any essentialist elements, showing that the predominant role assigned to this system and mode of production is the result of

---

22 The concept parasite in this text is used to describe, as biology does, a relationship between two or more elements, and does not have any political or moral implications.
23 Polanyi (1958) argues that the expansion of capitalism in Europe in the twentieth century was possible because of a planned destruction of the economic, moral and social order.

power relations, and of the complicity of political and academic fields. By dismantling the fictions around the hegemonic power of capitalism, it is possible to see how significant non-capitalist structures and relations of productions are, and thus to construct other truths about how the socio-economic order functions.

If, as Foucault argues, the connection between body and history is the place where genealogical analysis should be carried out (Foucault cited by Martinez-Novillo, 2010), it makes sense to attempt to dismantle the "certainties" built upon a market-oriented system by asking about its role inside the conditions that make the reproduction of bio-ecological, material and social life possible. This question is fundamental, particularly today when, despite evident global chaos, capitalism and its sponsors speak at length about their triumphs and the welfare they generate.

Through non-orthodox Marxist authors, the text has illustrated that capitalism is first and foremost a system of accumulation which at some particular point in history became a mode of production with particular forms to produce surplus value and control labor. This means that the wage-labor relation is not the principal feature of capitalism in the world-economy, but a circumstantial one, a situation that the actual crisis of capital accumulation would confirm. With this in mind, an important question to raise is how, when the certainty of full or partial employment is disappearing, are people managing to maintain the standards of living if it looks like capital no longer needs the wage-labor relations of production in order to accumulate.

The work of an author such as Gibson-Graham provides us some answers about the issue. Gibson-Graham (see e.g. 1996, 2001, 2002) has been working intensively with villages in Australia and the United States that suffered economic dislocation through privatization, deindustrialization and sectorial restructuring. The author witnessed how people within the core of capitalism must fortify preexisting structure-strategies or reinvent new ones to produce, distribute and consume the necessary goods to live and to rebuild the social order dismantled by the arrival and departure of capitalist investment. Through intensive research work, Gibson-Graham collected empirical evidence to show that in both "underdeveloped" capitalist societies as well as in the most "developed" ones, social organization is based on a complex network of non-capitalist structures that make individual and social reproduction possible.

Combining Marxism's tools with a feminist critique supported, in turn, by rich field work with local communities, Gibson-Graham proposes a practical guide to "taking back the economy". Her goal is to show, again, as Marx and Engels once did, that economy is the product of social relations and political decisions, and that the economy is what we (discursively and practically) make it (Gibson-Graham, 1996). The author de-naturalizes the idea about economy as essentially capitalist, destabilizing the hegemony of the dominant representation in all economic activities in terms of their relationship to capitalism – *as the same as*, *the opposite of*, *a complementary to*, and *contained within* capitalism. She develops a theoretical framework on i) the production of different representations of economic identity and ii) the different narratives of economic development. She reveals that the real world, where people are born, reproduce and die, is configured by a huge universe of economic relations, not just those oriented to capital accumulation.

If economy is what we make it, it would mean that an economy is not simply the structure that allows us to earn a wage, or to accumulate capital for maximum profit. Rather, an economy should be understood as the whole biological, material, social and symbolic processes that make us human beings, from the oxygen that we breathe as the ecologist movement insists, to reproduction, maternity and the household, as feminists have always claimed (Kollontai, 1921; Berk, 1979; Nelson, 1996, 2006; Folbre, 2008; Federici, 2012). Economic analysis, then, should include all the structures of production involved in human existence, such as the production of natural resources, the production of the human body and their social conscience, and the production of the elements of social order: values, norms, institutions, and technologies.

Communitarian exchange, self-employment, voluntary labor, household production, bartering, communal work, and so on are all active and functional structures of production and distribution that support daily conditions of human life in its material and affective dimension. Nevertheless, all these complex networks of environmental and social relations are neglected by those carrying out classical, and even critical, economic analysis. Such authors have ignored, for example, the reproductive capacity of the ecosystem to supply natural resources and the reproductive capacity of women to supply human laborers and consumers (Dunaway, 2001). Instead, economy has

45

been reduced to the production of means of accumulation and the production relations organized under the wage-labor wage scheme (Picchio, 2001). Capitalo-centrism has blocked our understanding of the real economic world, and the whole process that makes social and demographic reproduction possible has been displaced to guarantee capitalism takes an advantageous position not only within academic and political analysis and discourse, but above all within the distribution of the whole social surplus.

By analyzing relations between capitalism as a system of accumulation and as a mode of production and its connection with the real conditions of social reproduction[24], the cost-benefit of wage-employment relations would seem to play a very small role into the whole social production. Labor power, wages and employment have been tools implemented by capitalism to assume control over social reproduction, as well as of the wage-earning and unwaged labor power, exalting the former as productive citizens while portraying the latter as drains on society (Federici, 2013). Analyzing the role that capitalist structures and paradigms play in social life as a whole, one can see the huge distance between them and the real conditions for the reproduction of human life.

Capitalo-centrism has limited the economy to the production of commodities within the wage-labor scheme while eliminating all other schemata or structures of production. As a result, it has been easier to distribute, justify and hide exploitation, configuring, at the same time, the material and moral conditions so as to divide the exploited and force them into confrontations when it is necessary.

Feminist economic analysis has been engaging with this issue for decades by showing, for example, that this person arising out of thin air, i.e., "the worker", who in the economic literature appears ready to sell his labor force to the market does not exist. Correspondingly, neither employment nor wages are the core forms of human and social relations. Instead of "the worker", "the employed" and "the wages", there are males and females born on a planet that, free of charge, delivers them the biological conditions

---

24 Foucault talks about "bio-politics" to refer to the control of the biological process of the human body in society, particularly in the capitalist one (Foucault, 2008). But he disregards the previous process of production, caring and bringing up of the body.

to live in. Instead of the "wage-worker" there are human beings who depend longer and more deeply than any other creature on social attention and protection. Before a person becomes a "wage-worker", he or she has been supported by unwaged, non-proletariat structures of production, such as familial, communal and communitarian, and many of them are organized and supported by women.[25]

These familial, communal and communitarian structures are even more effective during the period a person becomes part of the wage-labor force. They provide the necessary social and biological conditions to keep the "worker" performing; they take care of children and households as well as of the physical and emotional health of the "worker". In addition, they give the "workers" care when they can no longer be employed because they are old, sick, or simply because, as we have seen during the last decades, the market does not need them anymore.

In addition, in places such as Tumaco-San Lorenzo, located on the periphery of capitalism, where expanded production has not been the basis of capital accumulation, but rather the exploitation of land, natural resources, and dispossession of people who "openly" receive much less than "the quantitative standardized monetary retribution" for their labor, these familial, communal and communitarian relations maintain personal and social life. To Black-African people living here, the reproduction of life has depended on the structures named above, but also on the environment produced by nature; land, fish, animals, farms, fruits and vegetables, trees, etc. We cannot forget that although in Tumaco and San Lorenzo capitalism had tried to take control of natural resources, there are still today, strong conflicts between native communities and capitalists, and despite the monetary, political, and military advances to support the latter's interests, black people have not faltered.

Nor can we forget that the environment is not a particular or exclusive condition of the support and production to Black-African communities; every human being depends on it for their own survival. Thus, although in core societies people do not guarantee their livelihood by fishing or farming their own food, nor by using native trees to build their houses,

---

25 One of the most recent works on these issues is <u>Who pays for the kids</u>, written by Nancy Folbre (2003).

without optimal environmental conditions human beings could not exist. It is therefore evident why, in the last few decades, environmental damage has become part of the global political agenda.

It can be seen that there is a complex net of structures producing conditions that make possible the bio-social reproduction of human beings. Based on information collected in the field as well as on some theoretical tools developed principally by new Marxist ecological thought and feminist political ecology (cf. Shiva, 1989; Hovorka, 2006; Elmhirst, 2011; Buechler & Hanson, 2015), these structures, a social metabolism, have at least four levels: i) environmental production; ii) production of human bodies; iii) production of social abilities and iv) the structure of production of material goods for consumption, exchange or trade.[26]

Classical, neoclassical and critical perspectives on the economy have limited their study to a small part of the whole process, i.e., the production of commodities via the wage-labor relationship. However, taking into account the whole process, the production of material goods within wage-labor relations, the production for obtaining and accumulating surplus appears as a very small part of a complex web of structures.

Wage-labor relations exist without doubt, nevertheless, considering the entire necessary structure of ecological, social, economic and even cultural reproduction, their role is smaller than we think, and has become even smaller over the last few decades. While the wage-labor relations within capitalism are becoming smaller, the rates of capital accumulation are increasing as never before, as well as the political power of capitalists.[27] The increase in global inequalities and the concentration of political power by capitalists as the chief consequence of the so-called economic crisis, as many argue, are the result of a monumental transformation of the ways in which capitalist accumulation works.

Nevertheless, changes and transformations have been constant factors within capitalism's history, as Arrighi (2000: 1) argues:

> [C]apitalist history is indeed in the midst of a decisive turning point, but the situation is not as unprecedented as it may appear at first sight. Long periods of

---

26 The point on the social metabolism is developed in the third chapter.
27 In 2007, the richest population quintile earned 83% of global income with just a single percentage point for those in the poorest quintile (UNICEF, 2013).

crisis, restructuring and reorganization, in short, of discontinuous change, have been far more typical of the history of the capitalist world-economy than those brief moments of generalized expansion along a definite developmental path like the one that occurred in the 1950s and 1960s. In the past, these long periods of discontinuous change ended in a reconstitution of the capitalist world economy on new and enlarged foundations.

Since the 1970s, the global political and military configuration has been changed dramatically to guarantee freedom of capital so as to increase profit rates (Harvey, 2003). During the last five decades, as in other periods, capitalism has been moving from concreteness, rigidity, and a narrowing down or closing of options to expanded liquidity, flexibility, and freedom of choice, as Arrighi (2000: 5) notes:

> Capitalist agencies do not invest money in particular input-output combinations, with all the attendant loss of flexibility and freedom of choice, as an end in itself. Rather, they do so as a means towards the end of securing an even greater flexibility and freedom of choice at some future point.

Capitalism does not have a particular relationship with specific commodities or productive activity. A structure is capitalist because of the fact that money is endowed with the "power of breeding" systematically and persistently, regardless of the nature of the particular commodities and activities that are, incidentally, the means (Braudel, 2006). Wage labor is a particular input-output, a necessary combination in the phases of the material expansion of monetary capital (Arrighi, 2000). But now, in the current phases of financial expansion and increasing mass of monetary capital, where capitalism "sets itself free" from its commodity form and accumulation proceeds through financial deals, wage labor relations have lost their relevance inside capital accumulation, and the workers have been seemingly freed, as Mazzadra (2013: 4) points out:

Once freed from the unilateral link with 'free' wage labor [...] the concept of labor power nicely accounts for the 'capture' of value produced by social cooperation outside the process of production, which more and more characterizes financial capital; and it further allows to theoretically grasp the spread of unpaid labor in the framework of contemporary precarization and flexibilization of labor.

We are witnessing a global process of "peripherization", characterized by the prevalence of unequal exchange through expanded production, where capital accumulation does not work through a capitalist mode of production characterized by wage-labor relations, but through the exploitation and dispossession of excess production under non-capitalist relations and structures of production. By this logic, what Mazzadra calls "flexibilization or precarization of labor" is, in the short term, the global expansion of the exploitation of labor for capital accumulation, but without the "guarantees" that, for a short period of time, in at least some countries, the wage-labor relationship of production provided[28].

Wage-labor as the core input-output within capital accumulation is vanishing, but people are not. When a wage labor position disappears or is "flexibilized", the affected person does not disappear, nor do the needs and responsibilities this person has (whatever they may be). But we need to remember that wage-labor is more than just the starting point of exploitation within the capitalist mode of production, it is the concrete result of its relationship with non-capitalist production.

Capitalism as a system of accumulation, throughout its entire history, has taken for its own benefit, rather than for the benefits of capital proprietors, the surplus of all non-capitalist structures of production. It has appropriated environmental resources such as precious stones, minerals, petroleum, seeds, animals, water, and has stolen human beings' skills to understand and transform the world. As many parasitic structures do, capitalism has taken the place of the real economic world while overtaking, at least in the discursive and analytical spheres, the non-capitalist structures that really support human life and constitute the true sources of capital profit.

Taking back the economy: Tumaco-San Lorenzo and diversity economic praxis

The phenomenon of a mass of "free workers" is a specific experience in specific societies within the core of capitalism. Throughout the rest of the world, in places such as the Tumaco-San Lorenzo region, where "full employment" of the labor force was never an issue, the relationship between capital accumulation and the work force has always been "flexible".

---

28 These guarantees are the product of hard won social rights and conquests of the workers movement.

This flexibility means that, for centuries, people have worked for capital while at the same time they have had to provide for themselves the necessary conditions for the reproduction of life and the labor force.

The relations of production, reproduction and capitalist accumulation in the regions studied could also help us to explain a dynamic around the world, where, despite the so-called economic "crisis", the social, biological and environmental world continues to reproduce itself. In the region of Tumaco-San Lorenzo the market-oriented system appropriates the excess produced by non-capitalist structures that allow and support the existence of environmental conditions and human bodies. The basis of capital accumulation is the ability these structures have to not only reproduce themselves, but also their resilience after every stage of dispossession, exploitation and plunder put forth by the capitalist system.

We witness how despite global warming, air pollution, the contamination of water, the destruction of forests, and the extinction of animal and plant species, the planet still provides us with the necessary conditions to exist. Women give birth to more and more children around the world, and these children are cared for and protected by familial and communal structures. The production of goods for self-consumption or trade has increased. All of this seems to be a paradox: the expansion of capitalist accumulation and primary accumulation has been possible because non-capitalist relations of production still exist. Yet it is anything but a paradox: Capitalism exists because it has been stealing, for centuries, the surplus from all other structures of production.[29]

Slowly the answer to the core question in the chapter about the conditions for new cycles of primary accumulation is emerging, as is the one about the conditions for the reproduction of life and the labor force in the periphery of capitalism. In Tumaco-San Lorenzo, both humans and nature have suffered many cycles of dispossession and exploitation on account of capitalist accumulation stretching back centuries. But in this space, people as well as nature have exhibited their resilience after every cycle of exploitation and dispossession. Production, dispossession and resilience constitute a

---

29 And it will continue in this manner until capitalism becomes a real structure of production i.e., until it is able to create all the conditions for its self-reproduction.

complex socio-economic cycle, which has not been understood as a whole, and therefore remains under-analyzed by scholars.

Some authors, however, have analyzed parts of this cycle. Norman, Jr. Whitten and Nina S de Friedemann (1974), for instance, have made attempts to understand the relationship between the capitalist system of accumulation and the non-capitalist structures of production in the location analyzed here. They say that the expansion and extension of the Black-Afro communities has been possible because of the development of an economy of subsistence, which allows them to overcome the effects of a fluctuating demand for raw material and natural resources. Another author who tried to approach the issue was Frey Rivera (1986), who argues that the basis of the social and economic organization in Black-Afro communities in Ecuador has been the ability of people to keep some independence and autonomy in the production of goods, thereby reducing the negative effects of the capitalist markets.

Arturo Escobar (2008) is an author who has worked intensively on the Tumaco region and the palm oil expansion. Escobar, who was in contact with the afro and indigenous communities on the Pacific coast of Colombia, has developed a complex scheme of categories such as "place", "capital", "nature", "development", "identity" and "networks" to explain, in his words, "the complex, historical and spatially grounded experience that is negotiated and enacted on every corner and region of the world, posing tremendous challenges to theory and politics alike" (2008: 1).

Escobar's (2008) research has been a great guide for approaching the complex reality of the Pacific region in Colombia. He has elaborated on many key issues not only to understand how capitalism works, but to explain how this system and the political and administrative praxis that support it have been challenged by people with other conceptions about social-economic-cultural relations while confronting the binary codes of modern-pre-modern or development-underdevelopment imposed by the "experts", whether they be scholars or politicians.

Despite these advances, there are gaps in the analysis and explanation about how, in the region of Tumaco-San Lorenzo, the whole circuit of socio-economic relations has been functioning throughout the last centuries. The analysis I propose is to study expropriation through capitalist accumulation, focusing on two areas: the functioning of those relations and structures of production based on non-capitalist rules that support human conditions for existence, and socio-ecological resilience.

This research partly follows in Escobar's footsteps, in an effort to dismantle the idea of capitalism as a powerful and hegemonic core socio-economic structure with the perspective that the economic world is ripe with many other types of practices and relations. But, unlike Escobar, who chose to ignore the footprint left by capitalist globalization in this region and how the region responds to it, this research intends to present, through a *longue durée* perspective, the principal elements, relations and structures that have configured the historical production, expropriation and resilience of this region and its people, delving into its most significant and ever more current example: the palm oil agribusiness.

Chapter II. The Long History of the Palm Oil Agribusiness and Development in Tumaco and San Lorenzo

In the last three decades, the exploitation of palm oil has turned into a central topic for activists, especially those concerned with the preservation of nature and human rights, but also academics. The relationship between palm oil and the destruction of the rain forest in Malaysia and Indonesia, the threats to orangutans and other species, the displacement and exploitation of native communities, particularly in peripheral countries, are topics of significant debate around the world. At this time there are recurrent global boycotts called for against companies that use palm oil. This situation gives us the impression that the production and commercialization of palm oil is something new; it looks like this commodity has suddenly emerged to draw our attention and disrupt our life.

This research attempts to move away from the "noise of daily affairs" in order to examine those elements in the early palm trade between Africa and Europe that could help us to understand the present reality of this industry. The aim is to show that the palm oil trade, and all of the effects this industry has had, are nothing new, and furthermore to expose how palm oil has for a long time played a key role in the configuration of the world as we now know it.

The production of palm oil and its sub-products involves many stages and components. This section focuses principally on the first stage of oil production that involves plantation, harvesting, collection and transportation of fruits, because this phase most clearly exposes the conflicts between Black-Afro communities, small farmers, businessmen, enterprises and governments.

Oil and fats are products best remembered by the dirty marks they usually leave; this chapter aims to pursue the marks left by the palm oil trade during the last two centuries. It is divided into four sections. The first section gives a general introduction to some botanical features of oil palms. The second analyzes the political, financial and scientific framework involved in palm oil production, and reviews the history of the palm oil trade from the fifteenth century to the end of the seventies in the previous century. By mixing the temporal and geographic expansion of the agribusiness, the chapter deals principally with the experience of West Africa, South Asia, and Central and South America, then focuses on Colombia and Ecuador. The third part analyzes the arrival and rise of the oil palm plantation in Tumaco and San Lorenzo from the 1960s to the present day, exposing all of the conditions which make the establishment of this agricultural industry possible in those areas.

The Elaeis guineensis[30]

The oil palm variety cultivated in Tumaco and San Lorenzo is the *Elaeis guineensis* (EG). It is a palm native to the west of Africa, in the so-called "Palm Belt"; Nigeria, Sierra Leone, the Ivory Coast, Angola, Gambia, Guinea, Ghana, Togo, Benin (previously Dehomey), Cameroon, the People's Republic of Congo and the Democratic Republic of Congo (formerly Zaire). This type of palm was cataloged in 1763 by Baron Nikolaus Joseph von Jacquin, and today the *Elaeis guineensis*, also known as African palm or simply oil palm, is the main resource for the production of vegetable oil around the world; it accounts for 36% of global vegetable oil. This palm has the highest oil-producing yield per hectare (Gupta, 2012), and in the last decades it has become one of the most important resources for the production of fuel from organic mass.

The *Elaeis guineensis* variety is a tropical palm which grows in warm weather, 84.200–91.000 °F max and 71.600–75.200 °F min, below altitudes of 500 meters. The EG needs 2.000mm. of rainfall, at least five hours sunlight daily or more, and can be cultivated in three different types of soil; loan, the optimal one because of its balanced concentration of sand, silt and clay; clay loam (more clay as the optimal variant), and sand-loam (more sand than is optimal variant).

Due to the aforementioned requirements, the areas on the Equator are the most suitable places to grow this plant. Nevertheless, for uncertain reasons, the wild variety of EG grows only in West and Central Africa,

---

30 The name *Elaeis guineensis* comes from the Greek *eleia*: olive, and *guineensis* from Guinea.

having been transplanted the world over. The EG is a monocotyledons plant without an authentic tree trunk, but instead a stipe that can grow between 25–30 m tall; in commercial yields, trees are replanted once the stipe reaches 12–15 meters. The palm has a complex structure of roots divided into three or four segments: i) the anchorage (more or less 10,000, 5–10 m thick, it gives support to the palm and extends to a depth of more than 50 meters; ii) the primary segments, which are between 2–5 mm thick and are not very long; iii) the secondary segments, 1–2 mm in diameter, and 15 cm long; and iv) the tertiary or smaller segments, which absorb huge amounts of water and nutrients (Syed, 2004; Ortiz, Olman, & Fernandez 2000).

Within this taxonomic classification EG is part of the monoecious group. Its female and male flowers have sexually specialized functions, and although they grow on the same plant, they bloom at a different period of time. The pollination process is cross-pollination; abiotic by wind, biotic entomophily where the pollination requires other organisms, in this case insect pollinators (mainly *Elaeidobuis* weevils) that carry the pollen grains from male flowers to female, and biotic anthropophily carried out by humans, often using artificial pollination and hybridization techniques (Syed, 2004; Ortiz et al, 2000).

Oil palm fruits are hard-shelled nuts which grow in large bunches, approx. 35 cm long and 50 cm wide. The oil palm seeds flourish inside the nuts, and in wild conditions they have problems germinating successfully because they require large quantities of oxygen and humidity. Due to this fact, on plantations, seeds are processed with oxygen, humidity and high temperatures (38.889–40.1°C) 80 days before being planted.

The oil palm fructifies between 24 to 30 months after sowing, and between five and six months after pollination, then bearing fruits every five or six days. Each fruit has an oily, fleshy outer layer, with a single seed also rich in oil. The bunch of fruit can weigh between 40 and 50 kg, and in general the fruit contains a percentage of oil of up to 50%, producing the highest yields per hectare of all oils crops at present (Corley & Tinker, 2003). Depending on the planting material, each palm can produce between eight to 15 fresh fruit bunches (FFB) per year weighing about 15 to 25 kg each.

The EG has been classified according to the variation of its fruits, seeds, and the size of the endocarp and mesocarp in three variations: dura, pisifera and the hybrid DxP tenera.[31] The commercial propagation of oil palm has been using the hybrid DxP seeds, which are obtained by controlled pollination, mainly from Deli dura or tenera variations. The seeds are cataloged depending on their geographical origin or the laboratory or company engaged in the selection (Corely & Thinker, 2000). In Southeast Asia a new variation, the Deli duras, a product of planting of seeds from the Congo in the early twenties, provided the basis to develop new planting materials used today by the industry in oil palm growing countries.[32]

The principal products for commercial trade are red palm oil from the fruit and kernel oil from the seed.[33] Today palm oil is extracted by a conventional, standardized and fully mechanical milling process, which demands a large investment of money and technology. After harvesting the bunch must be taken to the mill as soon as possible to avoid the acidification of the fruit. In the mill the fruit bunches are weighed, sterilized with steam at 140 °C for approximately 78 minutes. After that, the fruit is stripped of its fiber and nuts, digested and pressed to extract the crude palm oil (CPO), which will be then cleaned, clarified and stored. The process of producing kernel oil from the nuts is similar; nuts are separated from the fiber in the press and cracked, after which they are crushed in another plant to obtain crude palm kernel oil (CPKO).

Nearly 80% of palm oil produced is used in the food industry. Other products derived from the process are the palmist cake used as fertilizer or to produce food for livestock. The fruit and kernel oil are the basis for the manufacturing of products such as margarine, vegetable oil, cellulosic ethanol, soap, detergents, cosmetics, plastics, and varieties of fuels such as so-called "biodiesel" and napalm. Oil palm is today the most important vegetable oil for industrial production and has become the most highly commercialized oil around the world.

For more than a century, palm oil has been a part of daily life, not only in core industrial countries, but in global terms. In fact, nowadays, a large

---

31 The pisifera palms are predominantly female sterile and cannot be exploited for commercial planting. They are used for crossing with the dura palm to produce the tenera.
32 The development of Deli-Dura palm are recounted below.
33 The details of the oil palm production and the trade of oil are described below.

proportion of palm oil production is consumed in peripheral countries (Micha, 2014). A complex network of technological, economic and political power, which involves exploitation, social and environmental destruction and imperialist policies frames every aspect involved in oil palm production, from plantation, farming, oil extraction to commercialization. Making these structures visible is the aim of the following pages.

Oil palm expansion: A network of science, money and policy makers

A particular feature of human beings is their ability to alter, disturb and adapt the environment in order to create favorable conditions for their existence. By introducing animals and plants we have reconfigured the whole structure of the planet and determined not only the relationship between humans and nature, but between ourselves. But unlike previous moments of human history, when the reason for introducing plants and animals into a foreign environment was to produce food, clothing and tools or for medicinal or recreational purposes either for the clan or its leader(s), since the eighteenth century the introduction of species into foreign environments has had one and only one goal; obtaining monetary profit. The development of palm oil is a typical example.

The rise of the whole agricultural industry in general, and the palm oil industry in particular, cannot be understood if the economic, cultural and social context within which it has emerged is omitted. Modern monocultural agriculture, from sugar-cane plantations in the Caribbean and Brazil during colonial times[34] to the current agribusiness, have deep roots in three facts: i) the flourishing and development of mechanical philosophy since the

---

34 Sugar cane plantations were introduced in America during the sixteenth century for local consumption; cane was established as an export commodity in the mid-eighteenth century, principally in Brazil. During colonial times, the large plantations of sugar cane or tobacco were integrated into the Hacienda, a complex socio-cultural and economic structure whose goal was not only the production of commodities for the market, but the production of living conditions not only for the owner of the plantations and its large families, but also for slaves, craftsmen, priests, administrators and so on (for more information on mono agriculture in the New World, see Céspedes, 1983).

seventeenth century; ii) the Industrial Revolution and the implementation of the capitalist mode of production; and iii) the ascent of the bourgeoisie as the ruling class.

Despite its many variations, mechanical philosophy claims that the whole universe is a source of raw materials for the benefit of *Homo faber*. Within this conceptual framework, science and knowledge became tools to dominate nature – including other humans being – a tool accessible to just a few experts and industrious men, who at the same time were chosen to civilize and obtain the maximum benefit from people and nature (Alam, 1977). This philosophical, material and political spectrum produced networks of businessmen, scientists and policy makers, who, by working together, propelled a new stage of world conquest by Europeans.[35]

As was said above, wild *Elaeis guineensis* only grows in West Africa in "the palm belt". For this reason, capitalists have needed to find or produce ecological conditions to make the growth of groves according to international trade's expectations possible: high yields of fresh fruit bunches (FFB) with particular oleic acid levels.[36] As a consequence, the expansion of oil palm has been made possible through establishing a particular socio-ecological regimen, disturbing not only environmental but also socio-cultural structures in the places the crop is being planted.

The whole functioning of the oil palm agribusiness depends on the task undertaken by botanists and institutions such as universities, research centers and botanical gardens. These institutions have been key to the development of this kind of business, as E. J Corner (1966) noted in his book, <u>The Natural History of Palms</u>. Corner speaks of "the palm pioneers" – people and institutions who introduced the palms to the scientific world in Western societies. This group of pioneers included the Dutch, whose interests in botany were a matter of national interest. The Dutch equipped their universities with Botanic Gardens, building greenhouses for the cultivation

---

35 One of the most representative examples of this configuration was the Association for Promoting the Discovery of the Interior Parts of Africa, a British club comprised of members of London's upper-class, led by Sir Joseph Banks, a rich botanist and naturist, son of a wealthy family in Lincolnshire.
36 Subsidiary objectives are dwarf stature, for easy harvesting, genodermal resistances, tolerances to other diseases and stresses (Ngando-Ebongue et al, 2012; Corley & Tinker, 2003).

of tender plants[37], and to carry out the earliest publication describing the plants, their cultivation and uses.

It was precisely the Dutch who started the expansion of the oil palm by sending seeds from the Botanic Garden in Amsterdam to Asia. Then they built a Botanic Garden in Buitenzorg, now Bogor, in Indonesia, an institution that by this time was the main institution responsible for spreading the seed crossing the Atlantic and then to the Pacific Ocean. The garden was initially a private garden and mansion created by the Dutch East India Company, when in 1817 Caspar Georg Carl Reinwardt, a German-Dutch botanist, interested not only in obtaining knowledge but also in personal fame and fortune, officially established the Botanic Garden whose main task was the study of tropical plants. For a long time Buitenzorg was the institute responsible for sending oil not to new oil palm producers in Asia and Central-South America, but to former ones such as the Congo, Papua New Guinea and Cameroon (Corley & Tinker, 2003).[33]

It was no coincidence that Reinwardt headed the foundation the Botanical Gardens in Asia. He, as most botanists of his time, was a "plant merchant" or "agent of the empire" as Schiebinger (2004) noted, attempting not only to contribute to the wealth of the nation, but to benefit himself. This combination of interests can be deduced from Reinwardt's letter written in 1814 to the Dutch King Willem I, answering the latter's request to lead the investigation on the Dutch colony in the East:

> The available knowledge of the inhabitants, ... the land, and the great variety in nature of those countries [the Malay Archipelago] is too incomplete, further research would lead to new findings which would contribute to the enhancement of scientific knowledge in general as well as to the opening up of new exploitable resources, trade and prosperity ... The one who first takes up this career will probably already reap the rewards of important discoveries (Weber, 201: 15).

The figure of Georg Carl Reinwardt is emblematic but not unique in this constellation. As discussed previously, the spread of palm oil was possible

---

37 The term "tender" usually refers to perennial plants unable to withstand freezing temperatures.
38 Corley and Thinker (2003) developed a schema on the selection and breeding of oil palms concluding that many of the oil palm seed planted today in Asia as well in Central and South America come from the Botanic Garden in Buitenzorg.

thanks to the work of scientific and technical staff operating in three stages: i) in the research field, collecting data and information, analyzing the reactions of the palms in foreign environments, and adulterating the features of the palms to provide palms that met the requirements of the market; ii) in the ideological field, where scientific-academic knowledge went hand in hand with an army of experts to support and legitimize the concept of the environment as a source of raw materials and profit; and iii) as tools to control the access of people, in general, to the new technical developments and materials. For instance, in order to increase the oil content in fruits and kernels, public research institutions were opened to attend to the needs of palm oil producers. These institutions were responsible for improving, selecting and breeding seeds, as well as developing new technical advances in farming and pollination and introducing mechanical advances to the mills.[39]

Thus, without the work of botanists, chemists, engineers, geographers and anthropologists as well, the palm oil agribusiness would not have been possible. Other examples of the close relationship between oil palm expansion and scientists are the work of the chemist Nicholas Leblan, who in 1791 produced sodium carbonate from common salt, and Michael Eugene, who in 1800 discovered that oil and fats were a compound of fatty acids and glycerin. Both discoveries were key to the productions of soap, fire gum, and to the manufacture of stearin candles,[40] products that required palm oil for their manufacturing.

In the early 1900s another chemist, the German Wilhelm Norman, an example of the new profile of scientific and business men, introduced and patented the process of hydrogenation of vegetables fats, a process that transformed vegetable oils to make them malleable to industrial uses, especially the production of margarine and other manufactured foods. These discoveries, along with the development of the tinplate industry, meant a

---

39 The largest research institutes involved in the early research to improve selection and breeding of oil palm were The Institute National pour l'Etude Agronomique du Congo Belge (I.N.E.A.C) in Africa, the Allgemene Vereniging van Rubberplanters ter Oostkust van Sumatra in Indonesia, and the Department of Agriculture in Malaysia (Hartley, 1988).
40 Henderson and Osborne (2000) quote an old placard showing one of the founders of Price's company, a London producer of stearic candles, freeing a slave by burning away with a candle the rope which binds him.

radical transformation of food behavior in Western societies (Berger & Martin, 2000; Henderson & Osborne, 2000), consolidating one of the major industrial sectors demanding palm oil together with the cleaning industry.

But the work of the scientific staff is just a part of the whole chain; the other parts, which I now turn to, are the policy makers and of course the proprietors of capital. The next section discusses in detail the expansion of palm oil production, dealing principally with the experience of West Africa, South Asia and Central and South America, exploring the political and financial networks it involves.

History of the oil palm trade since the fifteenth century
West Africa

The palm oil trade with Western societies began in the fifteenth century, with the early colonization of West Africa. In 1482 the Portuguese founded a fort in Ghana and then reached Benin City in Nigeria, years later settling the coast of West Africa, where palm oil was very important in domestic requirements and commercial exchanges (Berger & Martin, 2000; Corely & Thinker, 2000). By the sixteenth century, with the establishment of the transatlantic slave trade, palm oil became an important article to supply the ships sailing across the Atlantic as well as to feed the enslaved populations.[41] For more than three centuries, the amounts of palm oil involved were negligible, palm oil imports continued to exist but as a relatively insignificant commodity for Western traders.

Then in the nineteenth century, the palm oil trade became a well-established business, and after the slave trade it became a secondary commodity in the West Coast trade, particularly to Britain, which increased palm oil importation from 2,233 cwt in 1802 to 1,058,789 in 1895 (Lynn, 1997).

---

41 Some authors argue that the traditional use of oil palm by black communities in Salvador da Bahia, Brazil, is evidence that oil palm was transported by enslaved populations from Africa to America, or by slave traders to feed them (Berger, & Martin, 2000). But due to the conditions of slave trade development, it is hard to imagine that slaves had a chance to bring food with them as if they were on vacations, or that dealers were worried about reproducing the diet of Africans in the New World.

Some explanations for this change are that by the nineteenth century, the panorama of international trade underwent dramatic changes; the slave trade remained the key export activity, but it was gradually dismantled, and after its abolishment in 1807 by Britain, the major slave trader in West Africa, it became a very risky and expensive activity. On the other hand, industrial production in northern countries rose, increasing the demand for raw materials to feed the growing industry, and palm oil was in this sense a valued product, used in the production of soap, candles, margarine, and glycerins, as well as lubricants for machinery and the tinplate industry (Corely & Thinker, 2000; Hartley, 1988, Henderson & Osborne, 2000; Lynn, 1997).[42] By 1850 oil palm was one of the main exports from Africa to Europe, particularly from the area known as the Oil River in Nigeria (Benin, Bonny, Calabar) and the former Dahomey Kingdom (currently Benin).

From the early trade in the fifteenth century until the 1870s, palm growers and oil production, as well as trade with Europeans, were controlled by local chiefs. This situation changed drastically after 1884 with the Berlin Conference, where the European powers, principally Great Britain, Belgium, Holland, Germany and France formalized the division of Africa, starting a fierce struggle to colonize and loot the continent and to expropriate raw materials; palm oil was one of the most valued articles of those times (Lugard, 1926; Lynn, 1997).

The division of Africa and the control of the production and trade in raw materials by European countries occurred in stages; first, by establishing protectorates in the former enclaves where the slave trade operated, then by penetrating and occupying the hinterland of West Africa, and in the end, by founding their own mercantile companies.[43]

---

42 Henderson and Osborne (2000) argue that the expansion of the palm oil trade was the main factor leading to the Industrial Revolution as well as the fall of the slave trade. Although it is always problematic to explain the development of a social phenomenon in terms of a sole cause or reason, as there surely exist many other factors in addition to palm oil, it did play a crucial role in the development of the Industrial Revolution and the end of the African slave trade. What is worth emphasizing is that palm oil has been a key element of trade and social life throughout the last five centuries.

43 In his book The Tentacles of Progress, Headrick (1999) explains that the plunder of Africa and Asia by Europeans after the mid-eighteenth century was only

By the end of the first decade of the twentieth century, the whole production of oil came from fruits that "local native communities" collected from wild or semi-wild groves, which grew in abundance. But due to the increase in the industrial demand for palm oil, the mere collection of wild fruits was no longer sufficient to meet the needs of the market. The problem for traders then became how to provide palm oil without increasing the cost of production. The solution was to establish monoculture operations.

A key step in the development of the modern oil palm plantations and its trade was taken by the English industrialists "Lever Brothers", founders of UNILEVER[44], one of the oldest and most powerful multinationals, producers of processed food and hygienic products around the world. In the early twentieth century the Lever Brothers controlled a large portion of the palm oil trade and production in West and Central Africa. Lever's search for raw materials in Africa began in the first decade of the last century, when William Lever intended to buy or lease land in the British West Colonies in order to expand the control of his enterprise over the production of palm oil, a key ingredient to Lever's soap industry. Initially Britain rejected Lever's proposal due to disagreements between British colonial government representatives and other British oil traders as well as the Nigerian authorities (Aghalino, 2000). Lever had to wait 10 years to take control of palm oil production in Niger and other British colonies as well the major oil producer in the world at that time; the Royal Niger Company (by 1930 the United Africa Company).

In the Belgian Congo the situation for the Lever Brothers was easier because the Belgian colonial government was looking for foreign investors to "develop" its territories. In 1891 William Lever and the Belgian Government signed a convention, or as Wilson (1954) described it, a "state document", which not only called into existence La Societé Anonyme des

---

possible thanks to a few but significant technical innovations such as the application of steam and iron to riverboats, improvements in firearms and the use of quinine prophylactic which reduced the death rate among European in the tropics.

44 Unilever is an Anglo-Dutch multinational resulting from the fusion between Lever Brothers and Margarine Unie in 1929. Margarine Unie owners were Jurgens and Van den Bergh, who also owned oil palm crops in Germany's colonies in Africa (to learn more about the history of Unilever and its role in the oil crop industry see Wubs, 2008; Wilson, 1954; Marchal, 2008; Fieldhouse, 1978).

Huileries du Congo Belge (HCB), but a contract to protect and defend its rights and interests through monetary, governmental and military means. This convention gave Lever rights over 750,000 hectares of land[45] divided into five circles around the whole country. Lever was allowed to collect native oil palm fruits or to plant palm trees where it was necessary, to build the necessary infrastructure for transportation, at the same time the Belgian Government would ensure that HCB had adequate access to land, labor and palm production by any means necessary (Fieldhouse, 1978; Wubs, 2008).

The convention imposed some obligations on the company, such as building infrastructure, roads, railways, one school and one hospital in each circle; these were easily fulfilled by Lever. Contradictions between Lever's and the native Congolese's land rights soon appeared. For instance, the convention ambiguously "regulated" the control and use of the land and legalized Lever's concession of Congolese territories, while it recognized "native" rights to collect and to commercialize oil palm fruits. The convention set out the conditions for labor management and its price; natives would collect and transport the fruit, receiving a minimum daily rate and food rations as remuneration, while Belgian citizens would be in charge of administrative and mercantile tasks. Then, it was decided upon that all native people living in these areas should work for HCB, cleaning and preparing the land, planting and seeding crops, fertilizing, fighting pests and diseases, and harvesting and collecting fruits.

Yet due to the fact that the natives were independent peasants and fishers and did not have any interest in working in Lever's fields, the colonial administration conceived and put into effect strategies to force them to do so. Supplementary taxes, such as head taxes or hut taxes, were implemented to monetize the local economy and force people to earn money. They were no longer allowed to pay taxes in kind, as was usual, they had to pay in cash. When a person or group could not pay taxes, which occurred ever more frequently, the men were forced to sign labor contracts to work on the plantations for a specified period of time. At the same time, Lever created "free" labor camps near the plantations where people worked and lived.

---

45 This extension of land is greater than the amount currently held by any oil palm plantation enterprise. Today the largest private palm-oil landholder, Sime Darby, has 529,705 hectares.

In these camps, workers used the money earned to pay the new taxes and if there were anything left after that, they would use it to buy food in the stores belonging to the company (Fieldhouse, 1978; Marchal, 2008).

In places where there were not enough youths or adult men able to work on plantations, HCB used slave laborers, many of whom were children (Fieldhouse, 1978; Marchal, 2008). Another strategy to "solve" the labor problem was to block communities' access to local markets. Independent peasants were forced by law to cultivate certain crops, including oil palm, cotton, and rice, which was in turn controlled by the colonial state; the latter managed everything from the distribution of seeds to the prices of the products. Palm oil made people in the Congo even more dependent on the colonial export economy than they were before, during King Leopold's reign (Houben & Seibert, 2013).

By establishing large plantation schemes to produce crops, Lever gave colonial relations another dimension and transformed, as never before, social relations and environmental structures in colonized places. Lever created new ways of accruing benefits from the colony by introducing state administration instead of private-personalized rule. This meant the introduction of new forms of cooperation between state and private interests, achieved by introducing the labor contract, and by "depersonalizing" and bureaucratizing power over the colonies. These reforms helped it to garner social and moral prestige by raising the flags of humanitarian modernization and progress to the Congo's people after the "scandal" regarding Leopold II's regime in the Congo.[46]

Lever's neocolonial scheme combined technological, political, economic and social reforms, introducing new labor control regimes (the wage-labor relationship), yet without completely disestablishing all of Leopold's regime;

---

46 The Congo, today the Democratic Republic of Congo, and between 1971 and 1997 the Republic of Zaire was a private colony of Leopold II, King of Belgium between 1885 and 1908. Thanks to the rights extended to him over the Congo in the Berlin Conference in 1884–86, Leopold became one of the richest monarchs in Europe by trading ivory and latex. Leopold's regimes of forced labor remain one of the most devastating genocides in modern times. Leopold's regime was responsible for the death of millions of persons in the Congo, (estimations range from two to 15 million) (for more on Leopold and his regime in the Congo, see Hochschild, 1998).

Lever was not all that different from the way in which Leopold administrated the Congo. Like his predecessor he created an industry that profited through dispossession, coercion and slave labor, as well as the mutilation of bodies (Marchal, 2008; Houben & Seibert, 2013; Fieldhouse, 1978).

Before the Second World War industrial production of oil palm production was concentrated in West and Central Africa, and in some areas of Asia. Except for the native communities in Africa, the main consumers of palm oil were industrialized countries. Between 1909 and 1913 the United States was estimated to have taken 20% of the world's supply. Within Europe, the United Kingdom was the main importer of palm oil, taking nearly half of Europe's supply, followed by Germany, Holland, Belgium, France, Italy and Portugal (Hartley, 1988).

After the Second World War a second boom in demand and palm oil production occurred. There was a new wave of industrialization in the core, but also in peripheral and semi-peripheral countries. This period saw the consolidation of multinational companies producing food (margarine, cooking fats, baking goods, ice cream) and cleaning products (soaps, detergents, pomades, glycerin) as well as the growth of chemical management in agricultural production. By that time, Africa was starting to be replaced by South Asia as the main provider of palm oil.

South Asia

After the Second World War the colonial territories of England, Belgium and France in West and Central Africa were still the major areas of palm oil production; nevertheless, the palm oil industry started to consolidate itself in the colonial territories of Malaysia and Indonesia. The first palm trees (Dura variety) in Southeast Asia were planted in 1848 in Buitenzorg (now Bogor) in Java, with seeds from Mauritius and from the Botanical Garden in Amsterdam. Some pioneers of oil palm as a commercial product were the Dane Aage Westenholz, head of the United Plantation Limited company, who early in the twentieth century already owned huge extensions of land and extensive latex extraction operations in Malaysia on the Sungei Bernam Estate, and the Belgian Adrien Hallet in Sumatra, a former worker of the palm oil industry in the Congo, who together with the Frenchman Henri Fauconnier were responsible for palm oil production in Malaysia. Both men established the first commercial oil palm plantations on the Tennamaram Estate to replace an unsuccessful coffee plantation project (Martin, 2003, 2006).

In 1918 the United Plantations company started small scale palm oil production by planting 2,428 hectares, which were located 60 kilometers up the Bernam River. Although oil palm production in Malaysia and Indonesia started under colonial occupation, it was only after gaining independence that the oil industry emerged as the key commercial product in both countries. The increase in demand for palm oil in industrialized countries after the Second World War, the drop in production in the Congo and Nigeria, and the collapse of rubber prices, the principal export resources in Malaysia and Indonesia, together with an accumulated knowhow in colonial agriculture, generated conditions which led the governments of both of those recently founded states to orient their agrarian production towards the palm oil sector.

The agricultural policies under independent governments did not harm the interests of the old colonial companies, and despite strong nationalist discourses (Donnithorne, Cortley & Allen, 2003), many old colonial companies continued to accrue significant benefits from the agriculture and commodity diversification policies of the newly independent states (White, 2004).[47] Once again Lever played a pioneering role by establishing an oil processing plant in Malaysia in 1952, developing breeding programs and introducing a process that pollinized the palm in a natural way (Cross, 1985).

Today, Indonesia and Malaysia are the largest palm oil producers in the world, with 9.4 and 4.6 million hectares respectively, followed by Thailand

---

47 In Malaysia, 70% of the oil palm planted was under private ownership, principally in the hands of companies with mixed national and international capital. But the continuity of the foreign companies is notorious, principally in the British-controlled palm oil industry. One example is the Sime Darby Berhad, the largest listed oil palm company in the world, which owns 531,229 hectares planted with oil palm, and controls 5% of the global oil palm market. The Sime Darby Berhad arose in 2007 from the fusion of Sime Darby, Guthrie and Golden Hope, three companies that from the early twentieth century have had land and rubber plantations and were pioneers in the introduction of oil palm crops in Malaysia. The company has been reported to have violated the land rights of local communities in Liberia, where they hold a concession for 220,000 hectares of farmland (for more information, see Pye & Bhattacharya, 2013; Friends of the Earth International, 2013).

with 644,000, Cambodia 118,000 and the Philippines with 46,604. In Malaysia and Indonesia, the governments have been committed to developing strategies to foster the palm oil industry by creating programs such as the Federal Land Consolidation Authority (FELDA) and the Federal Land Consolidation and Rehabilitation Authority (FELCRA) in order to guarantee the success of the industry. This is quite different to what happened in West Africa, where the plantations were driven by the promise of civilization and modernization. In South Asia, palm oil is framed within a discourse of progress, it is promoted as a tool to combat the high rates of rural poverty and create jobs for subsistence communities previously on or near the poverty line. Today, Malaysia and Indonesia produce 83.5% of the world's palm oil and 87.9% of the palm oil that is traded on the global market.[48]

In both countries palm oil plantations operate under three different but complementary growing models, i.e., private estates, smallholder estates and state-owned estates. Private estates with a single owner typically cover thousands of hectares of land densely and uniformly planted; 60% of plantations in Malaysia and 55% in Indonesia function under this scheme (MPOB, 2012). These estates produce palm oil through an industrial form of agriculture, requiring large investments of capital and access to the best agricultural and management practices. They are also involved in the breeding of seedling oil palms, the management of replanting schedules, the use of fertilizers and pesticides and pest management, labor recruitment, and integration with extraction operations (Colchester & Chao, 2013).

Smallholder estates constitute 35% and 40% of palm businesses respectively. The farmers on such estates have less than 50 hectares of land; Smallholder estates can be separated into two groups. The first is dependent smallholders, whose land is owned by individuals and who rely on large corporate private estates. They allow private estates to expand the planted areas of oil palm without making the investment in land themselves according to a predetermined, non-market price. The second group is independent smallholders, family-operated or larger estates operated with the help of

---

[48] Around 1995 Indonesia surpassed Malaysia both as the largest cultivator of oil palm by area and the largest producer of palm oil in the world because of the availability of lower-income workers and of land for plantation development, making the cost of production lower (Colchester & Chao, 2013).

hired labor. These plantations sell fruit on the open market, are typically less capital-intensive, and are highly sensitive to the fluctuating price of FFB. Yet, governments have promoted them as strategies for distributing wealth to lower-income households.

Government estates are government initiatives, principally in Malaysia, which are meant to encourage the widespread growth of oil palm cultivation and is promoted as a wealth distribution strategy for lower-income people. They operate with associated organizations of smallholders (Colchester & Chao, 2013).

Unfortunately, it was not possible for this research to find much available information on the early impacts of palm oil expansion in South Asia, especially on land distribution and labor relations. Nevertheless, the recent situation of environmental harm, child and forced labor in Malaysia, Indonesia, and other producer countries could help us draw a retrospective image of the past. Working in palm oil plantations is a very exhausting, difficult, dangerous and low-paying activity; correspondingly, it is primarily carried out by unskilled labor. In Malaysia, where there were agrarian policies in place that provided land and some autonomy to poor people through smallholder farming agreements, working in the palm sector is not desirable to the local population. Companies hire poor foreign marginalized migrants, principally from neighboring Indonesia, Bangladesh, India, Cambodia, and Thailand, vulnerable to labor and human rights abuses, and tend to be overworked and underpaid (Colchester & Chao, 2013).

Central and South America

As was mentioned previously, Latin America is home to a distinct native species of oil palm, in Colombia there are around eight of them (Ó Loingsigh, 2013: 23) however, they are not cultivated commercially because they produce very small fruit with a low poor quality oil content. *Elaeis guineensis* was introduced to Central and South America in the 1920s, but the growth and consolidation of the plantations and the industry has been slow and selective due to the large investment in capital and technology required in this line of business, as well as the recurrent land conflicts between peasants, landholders and governments (Richardson, 1995). Consequently, despite the fact that the first seed was planted in the late 1920s and the first commercial plantations began in the 1940s, it was not until the 1980s that the consolidation of oil palm as an agribusiness occurred in this region. This production is concentrated mainly in Colombia, Ecuador, Honduras and Costa Rica, but still does not represent a significant proportion of the global oil palm trade.

The small impact of its oil palm plantations and oil production in the global trade notwithstanding, the expansion and consolidation of the oil palm agribusiness in Central and South America has registered the same history of technical advances, political and economic interests as well as vigorous struggles as elsewhere in the world. Similar to what happened in West Africa and Asia, the introduction and expansion of oil palm was headed by a multinational company. In Central and South America, the United Fruit Company – formerly United Bread and today Chiquita Brands – was the counterpart of Lever. The first seed, Deli-Dura, was introduced in 1927, imported from the Botanic Garden of Buitenzorg Java to be planted in United Fruit's field in Panama. In the same year, United Fruit opened the Experimental Station and Botanical Garden in Lancetilla, Honduras, initiating the first program dedicated to the research and selection of oil palm seeds. That program would later be responsible for spreading the palm oil industry to other countries such as Panama, Cuba, Costa Rica, Guatemala and Colombia.

During the first 20 years, from 1920 to 1940, efforts in the palm oil sector were concentrated on experiments analyzing the behavior of seeds that came from different selection programs and places (Nigeria, Sierra Leone and Angola, the Congo, Malaysia and Indonesia, and the United States) in new environments, in addition to searching for potential producer countries.

The first commercial plantation was established in 1936–38 in Honduras by the Garcia family, landholders who planted 6.5 hectares. It was only in 1943 that United Fruit began the first large oil palm plantations in Honduras, working together and taking advantage of the experience and equipment of the Garcia family. By this time, the company also started to plant oil palm in Costa Rica as a substitute for banana production, which had been hit hard by disease.

For two decades United Fruit accumulated knowledge and experience developed by botanists and chemists in Asia, the USA and Central America.

Many former researchers at universities or government institutions later became administrative staff at the company, such as the botanist Wilson Popenoe, who for many years had worked for the United States Department of Agriculture and then became director of the Lancetilla Station for 14 years, and the chemist and professor W. G. Breck (Rosengarten, 1992).

The United Fruit Company's efforts yielded results after WWII, when most countries in Central and South America implemented import substitution policies in order to increase internal production and consumption and to also produce surpluses for the international market. Within this context, palm oil was promoted as a better strategy to reduce dependency on oil imports that countries such as Colombia suffered from,[49] and to further the establishment of industrial production.

Unlike Unilever, United Fruit did not face particular problems with regard to access to land in its bid to establish a large plantation scheme, since palms were cultivated on the former banana plantations the company owned. But the situation was different when it came to labor issues and the management of the plantation. This was due, on the one hand, to the fact that many workers came from the banana field and were already organized in trade unions, increasing labor costs and on the other hand, in spite of institutional and government support, there was neither enough investment in scientific research nor experienced administrators in those countries (Clare, 2012; Richardson, 1995).

The palm oil agribusiness only began to be successful in the late 1960s thanks to the rise in the consumption of edible oil on the continent, the consolidation of public and private programs for the management of seeds and plantations, and the development of structural conditions for the extraction and processing of oil. By this time United Fruit had established a network of control over the whole value chain of palm oil in Honduras and Costa Rica, giving it great power over it. United controlled the importation of seeds and also all of the knowledge on their scientific management, as well as the organization of the plantations, the administration of labor and its prices, the extraction process and the trade in oil. Lastly, it was the beneficiary of impressive yet very questionable political support (Clare, 2012).

---

49 By 1960 the consumption of palm oil in Colombia amounted to 70,000 tonnes, of which 41,000 were imported (Ridler, 1976).

Colombia and Ecuador

The history of the oil palm in Colombia and Ecuador is one of parallel developments. Although both countries were part of the international chain configuration set up by United Fruit in order to expand the palm oil industry, they went through a singular development. The first EG seed was introduced to Colombia in 1933 by the director of the Botanic Garden in Brussels, Florent Claes, who in turn used the seed from the Botanic Garden in Eala, Zaire (Congo). Some seeds were used as ornaments and others were sent to the experimental station in Palmira (Perez et al, 2003). But United Fruit established the first large commercial plantation in Colombia in 1945, in the Magdalena region, the heart of banana production in Colombia (Perez et al, 2003).

In Ecuador, United Fruit did not play a key role in the expansion of oil palm due to the fact that the Company used Ecuador to replace its banana plantations in Costa Rica and Honduras, which were in crisis as a result of the disease caused by the *Fusarium fungi* originating in Panama. The introduction of oil palm to Ecuador was the responsibility of botanist Lee Hines and the businessmen Leal and Roscoe Scott. Hines was part of the staff of the Bureau of Agricultural Economics, a branch of the US Department of Agriculture, in charge of the Technical Collaboration Branch, and director of La Estación Agricola del Ecuador. The Scott brothers, two US citizens residing in Ecuador, received Hines' seed that he imported from United Fruit's experimental station in Honduras.[50] The first plantations were located along the roadway between the provinces of Esmeraldas and Santo Domingo (Carrion & Cuvi, 1985).

As occurred in South Asia, the expansion and establishment of oil palm plantations and industry would not be possible without the support of the political administration, and the intervention of experts arriving from Asia, Europe, but mainly from the United States, as part of the "Alliance for Progress"[51] one of the most important international political and

---

50 See Organizacion de la Producción de Palma Africana en Ecuador (https://www.flacso.org.ec/biblio/catalog/resGet.php).
51 The Alliance for Progress, was not only a vehicle for the advancement of a new ruling class, but also the beginning of a new form of intervention on the part of the US government in the internal affairs of almost all countries in the region.

monetary initiatives of the time. Ecuador received special attention due to the introduction of an agrarian-export model, which was spurred on by the crisis in banana production and the ruling military dictatorship, while Colombia was propelled by the increase in social protest, and the formation of guerrilla forces. Against this backdrop, both countries designed import substitution policies in order to foment industrialization, build infrastructure, diversify production and increase internal consumption, as well as to reduce the influence of the leftist revolutionary movements amongst the farmers. The success of those policies was very limited in almost all areas, except in agribusiness, particularly the industrial production of some corns, which since then had experienced gradual but steady growth.

In 1957, the Colombian authorities initiated the Vegetable Oil Program by signing the Decreto 290, thus taking on the development of the palm oil business as a governmental program (Moll, 1987). This support for palm oil production occurred within a context in which the Colombian government pushed through land reform to placate huge protests and peasant demands.[52] The promise of land for the poor was, however, mostly rhetorical as large land tracts that should have been redistributed to poor farmers, ended up in the hands of powerful landholders, many of them palm oil producers such as Moris Gutt and Hipólito Pinto (Minga, 2013). Together with access to land, the Vegetable Oil Program also provided financial, research and technical support; during this time tax exemptions and subsidies were granted, along with policies to control the importation of oils. In addition, the main public institution in charge of agricultural issues started research

---

The governments of Colombia and Ecuador received credits provided by the AfP and administered by the Inter-American Development Bank under the supervision of USAID to conduct land reform, build infrastructure such as roads, ports and railways, and also received access to technical assistance. The arrival of the oil palm crop in Tumaco and San Lorenzo during the sixties was an AfP initiative, but its positioning in these areas in the last 15 years has taken place thanks to "Plan Colombia", a program of credit and military aid signed by the Colombian and US governments in 1998 (see La letra menunda del Plan Colombia. http://www.eltiempo.com/archivo/documento/MAM-1239857).

52 By this time Colombia was in the midst of one of its many bloody periods, especially due to the war between its main political parties, the Liberales and Conservadores.

programs on palm oil, the IFA, dissolved in the 1960s, and the ICA (Adams & Herron, 1965; Ridler, 1976; Vega, 1998, Quintero, 2010).

In 1961 Ecuador underwent a similar process by creating the "Programa Especial de Palma Africana", and a year later, the government, at the time a military junta, implemented credit policies, special research programs on oil palm, and also a state settlement program. The settlement program was an attempt to solve the growing demands for an agrarian reform. In particular poor farmers from the sierra were settled in the coastal territories where indigenous and Black-Afro communities were living, which resulted in a series of clashes and tensions (Gondard et al, 1983; Rivera et al, 1986).

Although some of the reform goals were to distribute land to the poorest and small farmers, it was palm oil producing companies, all of them private and some in the hands of military personnel, who received the best tracts of land. The argument behind this was that poor peasants were not in any condition to efficiently, rationally and profitably exploit resources (Cuvi & Carrion, 1985).

The production of palm oil has thus, from the beginning, been carried out by private actors. However, it would not have been possible or successful without the political administration of both countries. Governments encouraged financial and technical measures, as well as providing institutional support in order to guarantee the progress of the agricultural industry. In both Colombia and Ecuador early research programs on oil palm, as well as the training of technical staff and credits policies, were set out by the Alliance for Progress and the "Green Revolution", both programs supported and financed by international agencies such as the FAO[53], USAID and IADB (Aguilera, 2002; Adams & Herron, 1965; Ridler, 1976; Vega, 1998).

The oil palm was praised as a tool to reduce inequality and was promoted as a business accessible to everyone who wished to cultivate, but in the end the real beneficiaries of the administrative, financial and technical support were large landholders, who, at the same time, were powerful industrialists,

---

53 An example of the parallel development of the oil palm industry in both countries is visible in the person of Maurice Ferrand. Ferrand led the FAO's expert staff who visited Colombia and Ecuador and wrote reports on the environmental features of the countries, pushing the mass cultivation of the crop in this region.

relatives of politicians, or directly part of the government administration.[54] This network of businessmen and politicians created favorable conditions to increase the profits of the owners' plantation at the expense of the environment, as well as the physical and psychological health of workers, many of them small farmers who planted palm and had to sell their land to pay debts.[55]

By the mid-1970s the oil palm industry was not only a very well established and prosperous business in both countries, but the impetus behind one of the major transformations in the relationship between land, landholders, workers and rural communities the region has ever known. Large oil palm plantations introduced the modern agriculture-enterprise complex, displacing the "hacienda-finca" production relations that had been at the core of agricultural production since colonial times (Vargas, 2002). Palm oil enterprises have been under the control of some of the biggest industrialist-landholders, affiliated to FEDEPALMA in Colombia and ANCUPA in Ecuador. Both federations have been chaired by persons with very close ties to influential political circles and government administrations in each country.[56]

---

54 Ridler (1976) showed how by 1970 almost no small farmers in Colombia cultivated oil palm anymore because of the long gestation time the crop needs and the problems they had in gaining access to credit and technical support. In Ecuador the situation was similar: the colonization program accentuated the differences between small farmers and large landholders, due to the strong political influence of the latter. In short, public policies mainly benefited large oil palm producers, who not only maintained but expanded their plantations. Some emblematic cases are Ramón Pinto, owner of the company Gaseosas Hipinto, the Pumarejo and Dangond families, who to this day are two of the most powerful clans of politicians, landholders and businessmen in the country (see also Aguilera, 2002).

55 For a detailed report on the working conditions of the families employed on the oil palm plantations in Colombia see Minga, 2013.

56 Some examples of these close relationships are Jorge Ortiz Méndez, former chair of both IFA and ICA, promoter of the palm industry and associated with the company Monterrey S.A. The politician Antonio Guerra de la Espriella, who after working as chief executive at FEDEPALMA was appointed Minister of Agriculture. The current director of the company, Jens Mesa Dishington, considered one the most powerful men in the agricultural sector, is the husband of the Senator Maria Guerra de la Espriella, who is the sister of Antonio Guerra,

During the last two decades, oil palm plantations have experienced a renewed and accelerated expansion cycle in both countries, increasing from 114,561 hectares planted in 1990 to 403,684 in 2010 in Colombia (FEDEPALMA, 1994, 2013), and from 162,202 in the late nineties to 207,000 in the first decade of the twenty-first century in Ecuador. Today, with 403,684 hectares, a production of almost a million tonnes of oil per year, most of it to meet internal demand, and with possibilities for expansion to over three million hectares, Colombia ranks first in America and fifth today in the global rank of palm oil producing countries (FEDEPALMA, 1994). Ecuador ranks second in the continent and sixth worldwide with 244,393 hectares, and a production of approximately 500,000 tonnes of oil per year, 50% of which supplies the Colombian market.

The recent consolidation and expansion of oil palm in both countries has occurred principally in the region of the Pacific, and particularly in Tumaco and San Lorenzo. The arrival of the crop in this area has been framed by mass protests on the part of peasants, indigenous and Black-Afro communities. There are countless numbers of reports written by NGOs and government agencies on the links between palm oil and land grabbing, deforestation and pollution, displacement of communities, assassinations as well as coercion and exploitation. In addition, there are a significant number of palm oil entrepreneurs in prison because of their links with paramilitary structures. Suffice it to say that in Colombia and Ecuador, the expansion of palm oil agribusiness has been more the product of a perverse conjugation of political, monetary and military power, supported by coercive and criminal structures, than the result of successful entrepreneurship as is generally argued.[57]

---

and also the Minister for Agriculture in Colombia (2013–2104), Rubén Darío Lizarralde Montoya, who before heading the ministry, was the chief executive of FEDEPALMA.

57 There are many investigations and documents on this issue. Some used in this research include Minga, 2013; Soto, 2007; CECOIN, 2008; Comisión Intereclesiástica de Justicia y Paz, 2005; Mingorance, 2006; Defensoría del Pueblo, 2006; Documento de las organizaciones participantes de la mesa redonda de palma sostenible, 2007; Buitron, 2001.

History, development and features of the palm oil industry in Tumaco and San Lorenzo

In the previous section the research dealt with some "macro issues" involved in the oil palm agribusiness. It exposed the role of Latin America as a repository of raw materials and cheap labor. It also gave attention to the global and regional context within which the oil palm industry expands, analyzing some of the scientific-technological, political and financial threads that have been evolving over the last one and a half centuries in order to expand this particular type of agribusiness. In the next section, while maintaining this multidimensional perspective, the focus turns to the "micro level", exposing in detail how the whole structure involved in this agribusiness has been configured and how it has functioned in the Tumaco-San Lorenzo region.

As already mentioned, oil palm plantations are just one component of the whole palm oil industry; further components include the transport of the fruits to the mill, extraction, refining and commercialization of oil. The establishment of a plantation requires a large investment not only to obtain items directly related to cultivation, such as land, a labor force, seeds, water and the whole technical apparatus required, but to build the whole infrastructure needed to extract, refine, and transport crude palm oil, as well as commercialize it.

The oil palm trade is a complex and expensive business, one which requires conditions that Tumaco-San Lorenzo did not offer before the 1990s. Although the region was integrated centuries ago into the global market as a supplier of raw materials, it has not been the seat of an elite, nor was there a strong landowner class controlling agricultural production or a true capitalist class propelling the development of infrastructure and investing their capital in the classic sense. Quite the contrary, the traders and small merchants who arrived in the region left it soon after its more accessible resources were exhausted without establishing modern relations and conditions of productions. The elites, for their part, have mostly ignored the region and have scarcely generated development in a Western-modern-industrialized sense, as West (1956: 126) illustrates:

> Yet, despite the wealth extracted from gold- and platinum-bearing gravels, poverty has been the keynote of local economy for the last 300 years. Most of the lowland

> people still eke out a miserable existence through mere subsistence activities. Practically all of the Indians and most of the Negroes and mixed bloods are primitive farmers, fishermen and hunters, gaining just enough food to live by. Within the gold and platinum area, Negroes are small-scale miners as well as farmers and fishermen…Probably the only ones to gain substantial wealth from the lowlands have been owners and stockholders of mining concerns. Most of these people have been outsiders who were little concerned with the economic improvement of the area aside from their local interests.

However, in both Ecuador and Colombia political administrations are marked by a deeply racist hierarchy established during the early expansion of the colonial enterprise. The white-mestizo population assumed the role of a dominant-elite race controlling political, military and monetary power, while the rest, i.e. blacks and indigenous were seen as an "uncivilized", cheap and exploitable labor force (Quijano, 2000; Leal, 2008; Whitten, 1965). In an analogous manner, Tumaco-San Lorenzo is a region inhabited principally by Black-Afro communities descended from former slaves, whose livelihood is mostly subsistence agriculture, hunting and fishing. This has meant that since the establishment of the republic the region has remained, for all intents and purposes, marginalized from the political, administrative and business agendas of both governments. For centuries this area has constituted a kind of periphery of the periphery which the political and monetary elites have had an ambivalent relationship with; on the one hand, they see this area as savage, infertile and "unoccupied", but on the other, as a region with enormous natural resources and large contingents of potential labor, waiting to be exploited by "industrious and civilized men".

However, none of this means that the region has been outside the sphere of interest of businessmen, because, as was shown in the first chapter, this area has for a long time been a provider of gold, raw materials and food staples for the global market. In addition, it has served since the beginning of the palm oil business as an area of testing and exploration. It was here that, following the recommendations of the technical staff of the FAO, the most important research center and oil palm seed bank were built, i.e., the experimental station (Experimental El Mira) in Tumaco, and the experimental station (Forestal Agrícola Agroforestal) in San Lorenzo.

The arrival of the capitalist class, solving the infrastructure leak and the struggle against the "Negros"

Although the region of Tumaco-San Lorenzo still harbors small groups of white-mestizos, who constitute a kind of elite, a "true" capitalist class has never arisen to permanently rule the area. Here, the people doing business were sporadic merchants who were interested in raw materials, either plant-based, livestock or minerals, and after an initial boom they abandoned the area. The "true" capitalist class and landholder elite of both countries see the region as inhabited by ill and miserable "Negros", a place with an inhospitable environment, unworthy of any investment.[58]

The configuration of Tumaco and San Lorenzo as a/the core of the oil palm agribusiness has as its backdrop the concentration and monopoly of land ownership since the 1960s, a process that was accelerated in Latin America as a whole in the 1980s through the implementation of the neoliberal agenda and the commodification of the land,[59] as well as the consolidation of paramilitary armies, a phenomenon that is explained below.

The transformation of the role of land into a productive structure in a capitalist sense has also meant the consolidation of a "modernizing bourgeoisie" which, without abandoning their position as landholders,

---

58 In 2012, during a meeting of the departmental government, Rodrigo Mesa, a member of the regional parliament showed what many policy makers think about the region Choco-Pacífico and the people living there. The deputy said that to invest in Chocó was like trying to perfume or scent a turd (https://www.youtube.com/watch?v=Q0h3v_pZAGI).

59 Colombia and Ecuador have both since the 1960s been experiencing a huge wave of land concentration. Nevertheless, there are some manifest differences in the way in which this process occurred. For example, in Ecuador the state promoted a kind of agrarian reform to encourage settlement on "vacant land", mainly on the coast. But much of the land distributed was already occupied by Indians and Afro communities, who consequently lost their territories. On the other hand, the majority of new small farmers did not have access to either monetary resources or technology to produce and compete with the big producers. As a consequence, many small farmers had to sell their land to large landholders or abandon it altogether. In Colombia the state has failed repeatedly in the implementation of any policies to redistribute land. Here the concentration of land has been more the product of the absence of state action, and of the ability of landowners to expand their property in the absence of any state control.

diversified their influence by becoming at the same time industrialists and bankers.[60] This new modern ruling class is present in every major part of the productive sector and not only do they have political, monetary, military and technological advantages over black communities of fishers and hunters, they also boast a supposed higher moral standard supported by a Eurocentric and racist conception of Colombian and Ecuadorian society in which they see themselves as redeemers with a civilizing mission, as can be seen in the following text, written by H. J. Upeguí, former boss of FEDEPALMA:

> Nos metimos en canoa por el río Mira arriba, a buscar baldíos de la Nación, que más tarde nos adjudicaron. Nos ubicamos en el margen izquierdo del Mira, cerca de la frontera con el Ecuador, frente al poblado del Imbilí. Desde Tumaco se gastaba uno casi dos días para llegar allá. Era pura selva, no había nada más ... En la primera época a los nativos se les dificultaba amarrarse a un trabajo fijo. Uno los veía en la pobreza absoluta, pero ellos sentían que obtenían lo que necesitaban sin tener que hacer demasiado esfuerzo ... Recogian los cocos del suelo y comían coco. Se metían al río y sacaban pescado. Y tenían en el patio de sus casas una matas de plátano y yuca (Ó Loingsigh, 2013: 142).

Upegui's tale introduces us to the issue of what I call "the struggle against the Negros". In the sixties, during the first years of the introduction of the crop to Tumaco-San Lorenzo, the government created programs to promote the plantation within Black-Afro communities. Promising better living standards, the government tried to persuade farmers to cultivate the crop. However, just a few of them were seduced by the offer; for the rest, the crop was a kind of weed that would ruin those who cultivated it (Escobar, 2008). In this context, to persuade the communities, by any means necessary, was a key point.

In the first chapter I illustrated how the region had suffered many cycles of expropriation, showing how the oil palm agribusiness is just one more stage of the long history of exploitation within the process of capitalist accumulation. Every expropriation and dispossession cycle has left behind deep social and environmental impacts on the region and its communities. But it is after the second half of the twentieth century that what I would

---

60 Those events occurred parallel to each other in these countries and can be studied in the work of Abasalon Machado, Julio Silva Colmenares, Isabel Cuvi, Gondard, Pierre, Manuel Chiribocha and others.

call a systematic and strategic *blockade* of the native and Black-Afro communities living in the whole Pacific region occurs.[61]

Starting about sixty years ago, the Black-Afro and indigenous communities living in the Tumbes Choco-Magdalena on the Pacific coast between Panama to Ecuador started to be suffocated by land, sea and air. The land has been occupied and privatized by foreigners; oil spills[62] and other residues from the agroindustry have polluted the water, or it has been privatized by large fishing companies, and in the last decade glyphosate is now being sprayed not only over coca plantations but principally on subsistence crops.

This blockade has meant the communities loss of control over the land and other natural resources, as well as the loss of their autonomy over the production and distribution of the elements indispensable for the reproduction of social-material life. This has also meant the lack of independence regarding the administration of the labor force. Parallel to the blockade to obtain material elements for their reproduction, there has been a symbolic blockade altering the concept of living and working the land; today, for the new generations, being a farmer or a fisherman is considered low-prestige work which only unschooled and uncultured people do. The palm oil business has not been the only activity that has configured this blockade

---

61 In spite of having been pillaged for centuries, the whole region of Tumbes Choco-Magdalena, which extends from Panama to Ecuador, was scarcely explored and studied by the white man before the sixties. A rigorous exploration and inventory of its resources began in the sixties with task forces coordinated by regional administrations and supported by the FAO. An example of the latter are the texts Mision para el estudio y reconocmiento de los suelos de la región de San Lorenzo written by Alfredo Kupper and El noroccidente ecuatoriano (1959) and The forests of the northwest of Ecuador (1969) written by Misael Acosta, an Ecuadorian botanist who years later would be vice-president of the Alliance for Progress in Ecuador.

62 In 1996, for example, a fishermen's organization in Tumaco filed a claim for damage caused by oil spills as a consequence of oil transportation to Ecuador for almost two decades. San Lorenzo is cataloged as the area in Ecuador most affected by deforestation (for more on the demands for constitutional protection to prohibit the activities aimed at developing palm monocultures in the San Lorenzo region see Luna et al, 1999).

over the communities, but it is the one that has given it its biggest impulse and where most of the profits of this change have been accrued.⁶³

Arrival and establishment of palm plantations

The development of the oil palm agribusiness in the Tumaco-San Lorenzo region has taken place in phases; exploration-experimentation, establishment-consolidation and expansion. The stage of experimentation started in the sixties in the context of the import substitution policies and cooperation-intervention agreements. Ecuador and Colombia were visited by staff from the FAO and ORSTOM,⁶⁴ who analyzed the soil and climatic features of this region and defined the potential it would have to cultivate the crop. During this phase both experimental stations, Agroforestal in San Lorenzo and Mira in Tumaco, were built and the first "palm village", Tangereal, was established.

The exploration phase was a period without major conflicts between *palmeros*,⁶⁵ governments and communities. The main goal of this stage was to test the reaction and develop the crop in new environments. This phase was prolonged until the late 1970s, when the company Palmas de Tumaco built the first oil mill in 1977.

The second phase consisted in the establishment-consolidation of the industry. After being pressured for decades, with a population becoming desperate because their access to materials was *blocked,* the region was ready to receive the new "modernizing bourgeoisie". This phase began in the late 1980s and lasted until 2004, with a crucial turning point in the 1990s, where the conflicts inherent to capitalist expansion exploded. During

---

63 There is an intense dispute over the control of natural resources relating to wood, fisheries, and agricultural and mining businesses, not to mention that the drug trade has used the region as an enclave of production and commercialization of coca base paste. On the other hand, the area has been a focus of ideological and military dispute between guerrilla forces and paramilitary forces. The regional and international interests have also been shaped by the huge transportation, energy and telecommunications projects such as the Puebla Panama-Plan and the Initiative for the Integration of the Regional Infrastructure of South America (IIRSA).
64 Office de la recherche scientifique et technique outre-mer.
65 Companies owners of huge extensions of palm oil plantations and oil mills.

this decade the area became one the most important targets of capital investment, but also a place of open confrontation and resistance.

> Ellos (los palmeros) llegaron primero comprando por debajo del precio, pero no venían los dueños de las palmeras, sino que mandaban gente que se hacía pasar por pobre. La gente no estaba acostumbrada a ver todo esa plata, entonces vendieron. Pero luego, cuando la gente ya no quiso vender, porque se dieron cuenta que la plata se acababa y ya no tenían tierra ni nada, entonces empezaron a venir con amenazas, que si no vendíamos igual iban a negociar con la viuda. … Ahora sabemos que son de Colombia, ya no sólo son palmeras de Ecuador, sino de Colombia. Acá hay un poco de gente de Colombia que es dueña de las palmeras (San Lorenzo comunitarian leader. Interview. November 2012).

The golden era

In the 1980s, what the *palmeros* call "The golden era of palm" began (Guerra, 2002). This period featured aggressive projects dedicated to expanding control over land, natural resources and labor. Using the village Tangareal as a point of departure, the palm oil businessmen took over land declared "vacant" by the state. But this land was not "vacant", it was actually collective areas which provided food and raw materials to the Black-Afro and indigenous population. The palm companies began exerting direct pressure over those territories where people had their houses and farms, using three main strategies: first by using superior ties to the legal apparatus due to the fact that the majority of people did not have formalized land possession; then by shattering the farms and killing livestock, the people were forced to sell at very low prices; and finally by threatening and murdering all those who refused to sell to them.

The bloody era

In the 1990s the Black-Afro and indigenous population living in the whole Pacific region intervened actively in the political life of both countries. After months of social and political mobilization, the interests and counter-hegemonic vision and projects of Black-Afro communities were "recognized" as part of the nation, and the concept of *multi/plurinational states* emerged. In the period of 1991–1993 in Colombia and 1998 in Ecuador Black-Afro communities achieved constitutional recognition of the territories they had inhabited for centuries, as well as rights to control, administer and develop their territories according to their "ancestral" productive practices and collective interests. They also achieved some tools to fight racial-ethnic discrimination in order to re-signify their "Black-Afro identity". These events provided people on the Pacific coast with mainly legal instruments and "formal" institutional support to supposedly help them overcome the blockade they lived under, and theoretically to protect and defend them from the dispossession they suffered on account of many mining and agrarian enterprises.

The recognition of collective rights over the land inhabited for centuries by Black-Afro communities meant a small change within the power relations between palm companies and communities, challenging the absolute advance of capitalist interests up to that moment.[66] Now the states and their institutions have the task and the duty to protect and defend the interest of these groups.

Thus, while on the one hand, a given state recognizes and pretends to guarantee autonomy, control and administration of land, means of production and the labor force for the indigenous and Black-Afro population, it is renewing its support to capitalist agribusiness which is only successful by exploiting and mobilizing land, means of production and labor. Parallel to progressive constitutions that recognize the collective rights of indigenous and Black-Afro communities, the state reinforces its function as a spatial unity of the capitalist class in general. It operates as the administrator of the interest of hegemonic groups,[67] in this case, landholders-industrialists-bankers, promulgating new agrarian, monetary and infrastructure policies benefiting the oil palm sector in particular.[68] In Colombia, since 1998, the oil palm has been promoted through state policy as part of a program of coca crop substitution, within the context of Plan Colombia. Meanwhile, Ecuador terminated the road San Lorenzo-Ibarra in 1998, and in 2002 signed a law extending the agrarian frontier in San Lorenzo to 50,000 hectares at the expense of communities' rights, who lost approximately

---

66 Ley 70 (1993) in Colombia and Título III: De los Derechos, Garantías y Deberes/ Capítulo 5 in Ecuador.
67 The notion hegemonic group is taken from Poulantzas (1972), who understands the capitalist class as a plural and diverse group with a dominant sector who at the end controls the state according to its particular interests.
68 In 1983, the Colombian government passed a law on the deduction of liquid income and tax legislation in favor of specialized crops.

6,000 hectares and almost all the forest reserves, tending principally to the requests of the palm sector, which the family of the former Ecuadorian President (Gustavo Noboa Bejarano) is involved with.[69]

The constitutional changes that incorporated the interests of Black-African communities in the political arena did not mean a change in realities. The palm oil companies still had big advantages over the communities: They receive institutional support through credit and agricultural policies, and are part of the governments and their institutions (Guerra, 2002).

However, in spite of the power that the *palmeros* have, they still had problems getting enough land and an adequate labor force required for the plantations in San Lorenzo and Tumaco. And, to some extent, the situation worsened with the new constitutional framework and the new rights the communities achieved. The strategy used to solve the "problem" was a perverse combination of institutional support, which the palm businessmen have monopolized for decades, and para-institutional (paramilitary) armies, which have been legalized and reinforced by the Colombian government.

> Eso es como en los noventa. Yo me fui y cuando vuelvo está todo lleno de palma, hasta donde uno podía ver. Uno veía esto y la daban ganas de llorar. Eso fue una matazon, muchos eran paisas, con su carros, y sus motos. Pero no sólo era por la palma, mucha de esa gente está con la coca. Y empieza la matazon de gente, y las amenazas, y que son los paramilitares. Uno no sabía quienes eran esos, pero le decían auno que eran paramilitares. Esa gente llega para controlar la tierra, sacando y matando a la gente, y para asustarnos también, para que la gente trabaje para ellos, en la palma y en lo otro. Luego uno ve que los dueños de las palmeras son gente dizque muy importante de Bogotá, de Cali, de Medellín. Importantes? lo que son son unos bandidos. (Afro-Colombian Women leader, interviewed in 2012).

Paramilitary armies are the combination of state military forces with mercenaries. The role of this force in Colombia has been to take control over land, guarantee a sufficient supply of cheap labor not only for the growing coca production, but also for other agrarian activities, such as livestock breeding, banana plantations and, of course, palm oil (Soto, 2007). They have also had the task of preventing any kind of complaints and protests coming from the population.

---

69 The group Bejarano-Noboa is one of the most powerful business groups in Ecuador, and one of the biggest producers of palm oil.

Although the current incarnation of these armed groups arose in Colombia, in the last decade their range of action has been extended to the frontier, playing a key role in San Lorenzo, in the expansion of mining, saw milling and the palm oil industry, for instance, as well as in the control of the drug trade.[70]

Thus, contrary to the prevailing view on the paramilitaries as forces fighting against the state and counterinsurgency armies, or as an expression of "criminal capitalism", whatever this means,[71] the experience of the whole Pacific region in general, and Tumaco and San Lorenzo in particular, reveals that these armies are in fact part of a productive and political model. These armies have become very functional to the interest of elites in other countries (Azzellini, 2009; MOE, 2008), for example, in Ecuador. In the context of the Pacific region, the paramilitary structure must be understood as an instrument, firstly to counter the historical success achieved by Black-Afro communities within the political and constitutional arenas, and secondly

---

70 Of course, the development of a paramilitary force is not a phenomenon exclusive to Colombia, most countries have had similar schemes. But in Colombia, maybe because of the nonexistence of a true military dictatorship assuming the monopolistic control of violence and access to weapons as occurred in many other neighboring countries during the seventies, paramilitary forces have been growing fast since the 1980s, with legal, military and financial support from the regional and national governments. In 1994, former Colombian president Cesar Gaviria signed the Decreto 356 creating the "National program of cooperative neighborhood watch groups", legalizing in this way the first paramilitary forces named Convivir (Servicios Especiales de Vigilancia y Seguridad Privada). In the last decade these groups have expanded their control over land, and are very active on the frontiers of the Pacific territories between Colombia and Ecuador, without any significant reaction from the governments. It is also relevant that Colombia has become a kind of exporter of paramilitary forces, and in the last decade, this army, trained and armed in Colombia, has extended its activities to other countries such as Venezuela and Honduras (Azzellini, 2009).
71 Ivan Roa (2012), a researcher studying the palm oil industry in the same area I am, makes a distinction between legal and illegal. He says that the palm oil industry in this region is inserted into a network of criminal forms of capitalist reproduction. My argument is that being that capitalism is a world-system, that distinction is an analytical mistake. There is no legal or illegal capital, there is only capital searching for profit. Another thing is the moral justifications the capitalist used to control some aspects of the market. These moral justifications change with time, according to the modification of power relations.

to guarantee capitalists access to free or very cheap land, as well as to large contingents of cheap, intimidated and desperate labor.[72]

By 1996, the number of hectares under control of the palm companies in Tumaco increased from 14,000 to 30,000[73] (Escobar, 2008; Hoffman, 2004). The expansion of oil palm also means the expansion of large landholder properties and the reduction of collective smallholders. This situation has direct and significant impacts on the methods of land control and administration the communities have traditionally employed (Corponariño, 2008; Hoffman, 2004).

The bloody era continues

By 2004 many "problems" the palm companies had had in gaining access to land and labor had been resolved. The oil palm industry became the core of the entire productive activity in Tumaco-San Lorenzo, and the area was converted into the zone with the quickest expansion rate of oil palm plantations in the last decade, totaling more than 50,000 hectares (35,000 Tumaco and 17,000 San Lorenzo). Meanwhile, there was an increase in the number of those working for the *palmeras*, and likewise in the number of smallholders growing the crop for companies under the "productive alliance" model.[74]

---

72 In November 2014 rulings against the seven main heads of the paramilitary army, by the Tribunal Sala de Justicia y Paz del Tribunal de Bogotá, the magistrate confirmed the fact that the paramilitary structure has as its principal task to control and intimidate the population and not to combat the insurgency (see Tribunal Superior del Distrito Especial de Bogotá, Sala de Justicia y Paz, Magistrada Poenente Alexandra Valencia Molina, 2014).

73 Not all 30,000 hectares were planted with the crop, but were nevertheless the property of large palm companies.

74 A "Productive Alliance" is a model of association in which the government, large agribusiness companies and small farmers take part. The Alliance maintains the legal (not operational) independence of the different companies or links in the chain, and is formalized through contractual obligations. The small farmers provide the land and labor force in order to produce the specific stock the company requires and receive in turn monetary and technical support provided by the state. Under this model the companies control the price of the stock and reduce the costs and risks of production, which fall on the shoulders of smallholders. About a quarter of Colombia's palm is now cultivated by small farmers thanks

The establishment-consolidation phase of the oil palm agribusiness (1980–2004) left a balance of more than 16,000 hectares of deforested primordial forest, the pollution of the region's main rivers (Mira, Mataje Caunapí, Bogotá) either by chemicals products used to fertilize and to control diseases or by oil and grease waste. Other water sources were simply dried up. There is a long list of wildlife that no longer exists, and thousands of people have been poisoned, displaced, threatened, and hundreds more murdered.[75]

After fifteen years of brutal violence, forced displacement and threats to the Black-Afro communities living on both sides of the border, companies established and consolidated palm oil in the region, and the production of the crop became the core of productive activities in this area. Since 2006, Tumaco and San Lorenzo are again receivers of new governmental policies to cultivate palm, this time framed by a growing global demand for biodiesel.[76]

The costs paid by the inhabitants and the environment, the loss of territories, murders, and the pollution are silenced by governments and companies, but also through cooperation with institutions and even academics. For them, these "damages" are not part of the nature of agribusinesses themselves, but simply "irregularities". They even argue that malicious and misleading information has been spread about the oil palm agribusiness with the aim of putting society at large against the palm sector. They also say that the allegations of human and labor rights violations are mainly

---

to government loan incentives. Farmers have a three-year period of grace to begin repaying their loans. As a result, farmers feel they are becoming exclusively dependent on cultivating palm as a cash crop, and are often compelled to sell their fruit to the companies at below market prices.

75 These issues are dealt with in detail below (further information available in Altropico, n.d.; Restrepo & Cortez, n.d.; Mingorance, 2006; Buitron, 2001; Potter, 2013; Batallas, 2013; Magistrado de la Corte Constitucional Manuel José Cepeda, 2004).

76 In a meeting with the major palm oil companies in Colombia in 2006, the former president of Colombia, Alvaro Uribe Velez, reaffirmed the commitment of the state to the development of the palm oil sector, also establishing Tumaco as the focus of new investments and urging the black-afro communities to get involved in the agribusiness. In turn, the Ecuadorian government introduced a new National Biodiesel Program.

ideologically motivated in order to discredit palm oil as a sustainable commodity on the international markets, while any talk of a symbiotic relationship between oil palm expansion and the paramilitaries is flatly denied (FEDEPALMA, 2006; Rangel et al, 2009).

In the official discourse, the large list of businessmen charged with land-grabbing, murder, having links to the paramilitary army or practicing corruption are isolated cases, isolated actions of a small number of criminals that have intimidated even oil palm companies (Quevedo & Laverde, 2008). For governments, companies and development agencies, the "problems" associated with the oil palm plantations will be solved solely through so-called RSPOs (Roundtable on Sustainable Palm Oil) certifications.

Inside palm oil plantations

Palm oil plantations have transformed a large part of a region rich in biodiversity into a sterile oil palm landscape, where only rats and flies breed. As well, they have changed the land tenure relations and the modes of control and administration of work while disturbing the ancestral brotherhood of the two villages. In Tumaco, 44,000 inhabitants out of a total of 160,000 are working for the palm industry, but only 3,500 of them are permanent workers. In San Lorenzo, with its 45,000 inhabitants, approximately 5,000 work on the plantations, but less than 2,000 of them were hired as direct employees.

Palm oil plantations around the world are constantly denounced because of exploitative practices and underpaid workers. The plantations in Tumaco-San Lorenzo are no exception:

> El trabajo en la plantación empieza a las 6, pero la gente tiene que estar lista como desde la 3 para alcanzar el camión que los lleva a las palmeras. (…) Da pena ver como llevan a la gente, como bestias, una encima de la otra. Cuando llegan allá, (a la plantación) desayunan a la carrera, comida toda fría, muchos tienen problemas de estómago y hasta cáncer. (…) Los tienen como esclavos, trabajando como en la época de la esclavitud, con el sol quemándonos la espalda, y cargando pepas como mula (Former plantation worker, interviewed in San Lorenzo, 2013).

The majority of the population coming from this region who work for the plantations are hired under an outsourced model to do "semiskilled" or "unskilled", low-paid activities, including but not limited to conditioning soils and seedling, pest controls and fertilizing, pruning, collecting and transporting the fruit. By contrast, white-mestizo people from other regions perform administrative and scientific tasks. The supply of "unskilled" workers operates under three principal variants: i) An external contractor hires people and pays them a salary for fixed periods of time to do some of the tasks described above; ii) Short-time contracts are made with the company directly, mainly for collecting bunches; and iii) A local contractor, who in general is an inhabitant of the area, gathers some of his relatives or fellow community members to work the palm fields.

The working day in the plantations starts at 06:00 and finishes around 14:00. Nevertheless, to the workers, especially to women, the working day begins earlier, around 03:00 because they are responsible for household tasks. A huge number of workers live in remote areas accessible only by canoe or boat. They need between one and a half to two hours to get to the place where a truck designed to transport livestock picks them up at 05:30 to bring them to the plantations.

The daily salary earned by a worker performing "unskilled" tasks is calculated on a piecework basis, and is based on the kind of task they perform rather than who is the employer. The fruit is collected by teams made up of two or three people: one middle-aged man (*el cortero*) cutting the bunch, a gatherer (*papera-mulera*) collecting fallen fruits (usually a women or young man), and the driver. The salary for the group is calculated based on how many tonnes they have collected monthly. The price per tonne varies from seven to eight USD, of which the *cortero* receives 70%, since he does the most dangerous and exhausting work. The average amount of fruit collected by a group oscillates between one and a half and two tonnes per day, which means the group earns 14–16 USD per work day. Jointly, they make between 230 and 300 USD per month, of which the women and young men receive 100 USD or less.

Some workers hired by foreign contractors or "directly" by the companies earn part of their salary through a bonus that can be used only in the companies' stores (*comisariatos*). In these stores the price of the goods (mostly food and appliances) are higher than in other places, but due to the fact that the bonus can only be used in those stores, workers do not have much choice. In addition, in these stores workers have access to "credit", thus they can obtain products and pay for them later. Those hired by the local contractors must often work for their relatives to pay off favors or obligations, configuring an endless cycle of debt from which the worker cannot escape.

Many people working in the *palmeras* have suffered severe mutilations, others have become sick or even died because of the working conditions in the plantations and the frequent exposure to and mismanagement of pesticides. Nuñez's (2007: 123) research expounds on these issues[77]:

> Los trabajadores presentan permanentemente síntomas de envenenamiento como la boca amarga, mareos, náuseas, sofocación, debilitamiento, manchas y granos rojos y despellejamiento de las manos. Adicionalmente, la falta de control causa que los trabajadores tomen los productos libremente y los lleven a sus hogares poniendo en riesgo a sus familias. De los 33 casos atendidos en el hospital público (Hospital Divina Providencia, San Lorenzo) en el 2003 hasta mayo del 2004, el 67% corresponden a intoxicaciones de niños y 9% a muertes por accidentes en plantaciones y suicidios. Sin embargo, el número de casos sería mayor puesto que los intoxicados también acuden al hospital del Seguro Social y a doctores particulares, y toman remedios caseros ante intoxicaciones leves.[78]

Workers do not have effective collective bargaining rights, although in San Lorenzo there is a "trade union",[79] and the social security system is deficient. Workers commonly complain that companies either delay or do not pay their contributions to the health system at all, despite the fact that workers pay monthly contributions to have access to social security. This obviously leaves workers unprotected in the case of illness and accidents. Women have no special rights or considerations, and if a woman is pregnant she is fired without any compensation. There are also claims of repression through official or para-official forces on direct orders from the palm companies.[80] In addition, the companies bring in and hire people from other regions as a strategy not only to lower costs – as Colombians working in San Lorenzo earn much less and must work harder – but also to pressurise

---

77 This situation was a recurrent theme during the interviews, but unfortunately, there is not much independent research on these issues.
78 For more on this topic, see Movimiento Mundial por los Bosques, 2007.
79 After a strong confrontation between the communities and the companies, the companies created a trade union which, according to the communities, does not represent the interests of the workers.
80 This situation has been documented by the community in San Lorenzo, in <u>Conflictos Socioambientales en la Comunidad de Carandolet Relacionados con la Actividad Palmicultura.</u> (Cañas, 2009)

and intimidate the workers, thus fueling hostility between people on both sides of the border.

Working in the palm plantations as a wage laborer is just one mode of engagement with the agribusiness. There are also small farmers associated with large enterprises. Those producers are cataloged by the government, companies and cooperation institutions as "partners". They are really smallholders growing from one to 10 hectares. Under the "Strategic Alliance", supported by the USA, EU governments and cooperation agencies, small farmers receive loans and basic technological instruction and are obliged to only plant oil palm on their farms and to sell the produce to a particular company.[81]

As was stated previously, the oil palm is not a new crop to farmers; many of them started producing it in the seventies together with subsistence and other commercial crops. However, oil palm planted under the "Strategic Alliance" model has broken up this more pluralistic approach to land:

> (La gente) parecían locos tumbando las matas de plátano, de naranja, yuca, chocolate para sembrar palma. Casi todos por acá echaron abajo toda mata que tenían, no importaba de que era, lo importante era hacer espacio para sembrar palma. Ahora muchos estamos arrepentidos porque por la PC nos quedamos sin nada, hay algunos que ya ni siquiera tienen un verde[82] para la sopa (Farmer of Tumaco. Interview. December, 2014).

The first "Strategic Alliance" was established in Tumaco by CORDEAGROPAZ, a public-private entity founded and headed by the most powerful palm oil companies operating in Tumaco, in association with government institutions.[83] Since then, smallholders have created their own cooperatives or associations in order to "balance" the power relationship between small producers and large companies with regard to the prices for a tonne of palm oil fruit as well as for fertilizers and pesticides.

---

81 This model is in effect mainly in Tumaco, but is being promoted in San Lorenzo as well. During my field research, the Asociación de Pequeños Palmeros was expecting to receive loans from the government to cultivate oil palm.
82 Local name of a plant used in cooking.
83 Astorga, Palmas de Tumaco, Araki, Palmeiras Colombia, Santa Elena, Santa Fe, Palmas Salamanca, Agrigan, Inversiones Rankin e Inversiones Blum (see FEDEPALMA, n.d.).

After the production costs are deducted (monthly fees to pay off loans, seeds, fertilizers and pesticides, as well as wages) a smallholder can earn between 500 and 800 USD monthly. In comparison with other cash crops they used to cultivate, oil palm has meant a significant rise in earnings for small producers compared to other legal practices. Between 20% and 40% of the whole oil palm production in Tumaco-San Lorenzo is cultivated by small farmers (Escobar, 2008; Potter, 2013; Buitron, 2001; FEDEPALMA, 2013; Burgos, 2013). These farmers mainly produce under a family labor structure, or by contracting local workers, usually for harvesting work.

The "Productive Alliances", as Escobar (2008) foresaw, transformed the productive relations within the communities. Many farmers abandoned the subsistence crops and others they used to cultivate for the market such as cocao and banana, peach-palms, borojo and others to grow only oil palm.

Small farmers feel that they became highly dependent, probably as never before, on cash payments to support their families. They also think they are becoming dependent on chemical products such as fertilizers and pesticides and technologies which they have no control over. As a consequence, small farmers have lost autonomy over the management of their land, while at the same time alienating themselves from the natural and social context they once lived in. This productive model has also made them the most vulnerable element in the chain of production. They take on the whole cost of production while the companies get a guaranteed source of fruits without assuming any risks. Thus, any changes affecting the crops and especially the international oil markets will affect first and foremost the small producers.

This was the case in 2008 when the global price of oil dropped (ANCUPA, 2009), or in 2009 when the Bud Rot disease attacked the fields, destroying thousands of hectares, including almost 90% of the oil palm in Tumaco, and 50% in San Lorenzo. What is more, the "Strategic Alliance" model has fueled division within the population, bringing different conceptions and projects regarding the land, collective life, and the use of nature to the surface. One can distinguish confrontations between three groups: those pleading for individual ownership of the land and the possibility to sell it freely, those arguing for collective property with the possibility of growing oil palm or whatever the market has a demand for, and those defending collective property and conservation in order to, first of all, guarantee people's access to food and raw materials.

However, after analyzing the interviews and other materials collected during the field research, it is clear that there are no significant contradictions in the conceptions that the people in the rural areas of Tumaco and San Lorenzo have of land and territories. Maybe because of the political mobilization initiated in the 1990s, the local population has developed and strengthened a sense of community and belonging; in other words, perhaps the conflicts people are having are a direct response to the growing blockade they have been living under for centuries. Yet the situation is by no means clear cut. While for many of the region's inhabitants the oil palm deepens poverty and destruction, for others the crop means the best *subsistence alternative* they have ever grown, as seen in the following testimony:

> A mi me gusta mucho el cultivo. Con él (la palma de aceite) usted sabe que tiene todos los meses un ingreso, y que no tiene que esperar y humillarse para alguien quiera comprarle la cosecha como pasaba con el cacao. Con la palma uno tiene seguridad, ... y si uno va a la tienda a pedir que le fien y que al final del mes paga, pues le fian porque saben que de verdad al final de meses uno tiene su plata (Interview in Tumaco, 2012).

Nevertheless, by analyzing the concrete situation of most inhabitants of the region, the progress the oil palm has brought to small farmers is more rhetorical than reality. First, the profit from the palm oil business is accrued mainly through processing the oil, not planting the crop. Small farmers know this, and for that reason they are planning to collectively buy an oil mill. Second, the credit policies promoted by the government have not been accessible to the small producers, and when they do get access to some credit, it quickly becomes a burden they cannot bear as the farmers start to produce only to pay off the loans, which in some cases involves 14 years of installments.[84] On the other hand, cash income has not improved living conditions of farmers and their families. The monthly income for a smallholder oscillates between 500 and 800 USD, an amount that has allowed some of them to improve their houses by replacing wooden structures with cement ones, and to have access to some appliances and services such as motorbikes, cable TV, or better work tools. But people's basic living

---

84 Due to the Bud Rot disease, the situation has become even more complex since 2009. This disease bankrupted every small producer, forcing them to seek new loans.

conditions have barely been transformed: there is no fresh or clean water and no sewerage system, nor is there a permanent electricity supply or waste collection; education and health systems are precarious; and the infrastructure that people regularly use is limited to dusty roads. Despite promises of prosperity and welfare, the palm companies are in fact immersing the region of San Lorenzo-Tumaco in a social and environmental crisis. This has become a "Red Zone", with the highest rate of poverty, criminal violence and exclusion compared with other regions in both countries.

Conclusions

The development of the palm oil agribusiness in Africa, Asia and Latin America has occurred in a cyclical movement within two interconnected phases: exploration-expansion and establishment-consolidation. During the exploration-expansion phase, the efforts of the palm businesses are focused on research issues: the analysis of seeds, plant development in foreign environments, and the study of the environmental and social, economic and political characteristics of places where the palm could be planted. This is a long phase that could take decades since it involves complex research issues, and especially the creation of favorable environmental and social regimes guaranteeing the profitably of the business. This phase implies a huge transfer of resources, not only of seeds, but also of institutions, knowledge and money at local, national, regional, continental and transcontinental levels.

The first phase concludes successfully with the establishment of the plantation and the structural imperative to make it profitable. This structure includes the availability of enough land and labor, access to roads for the transport of bunches, a mill not far from the plantations to process the fruit, and the support of the given political administration. The plantations are established, generally, under the following schemes: large private estates, dependent or independent smallholders, and state-owned estates. Once consolidated, the plantations begin with the identification of new areas to spread the oil palm plantations, thus starting a new phase of exploration-expansion. The cyclical movement of the oil palm business has conquered almost all tropical lands around the planet, and a coordinated network of traders, scientists, mainly botanists and engineers, as well as policy makers have made this possible.

Historically, the palm oil agribusiness was expanded-established and spread in three phases. The first phase occurred between 1840–1920, due to the consolidation of the Industrial Revolution in Europe, the race for conquering the international market and raw materials, and the global deployment of industrial advances. The second phase (1950–1980) had as a backdrop the bipolar configuration of the global order with the strong intervention of the hegemonic blocks, principally the USA in Latin America and Southeast Asia, and the end of old-classical colonial structures. This was followed by the rise of independent and nationalist governments, and the establishment of industrial production outside Europe and the USA. The last expansion phase, when the region of Tumaco-San Lorenzo was contemplated, started in the late 1980s. This phase was framed by an accumulation crisis of capital, and the consolidation of financial and corporate capitalism with a neoliberal agenda. This new stage of the capitalist system has mainly generated, on the one hand, the deregulation of wage-labor relations, principally in Europe and the USA and on the other hand, it has renewed a violent process of privatization, commodification and monopolistic control not only over the means of production in the classical sense, but also over knowledge, information and natural resources such as land, water and seeds.

The expansion and establishment of the palm oil agribusiness in Tumaco and San Lorenzo has implied a deep dislocation of the social and environmental conditions that have supported the existence of the Black-Afro population over centuries. This dislocation has occurred not only through the dispossession of the material conditions necessary for social reproduction, by subjecting the people to produce for others, but also by advancing an ideological project in which life is centered around the need to obtain monetary profit. Both processes, material dispossession and ideological colonization, are framed by the intrinsically imperialist character of capitalist expansion.

Chapter III. Imperialism, Unequal Exchange and Palm Oil

> *"I am asked where is Imperialism? Just look into your plates: you see imported rice, corn or millet. This is imperialism. Let's not look any further"*
> Thomas Sankara.

When analyzing the evolution of the palm oil business in West Africa and its further expansion to Malaysia and Indonesia, it is hard to deny the imperialist structure and development of this business. However, in revising academic literature on imperialism today, it is in fact difficult to find someone, besides activists and some leftist academics, arguing that the current situation in the Pacific coast of Colombia and Ecuador, or in Asia, is a product of imperialist relations or is framed by them, and thus the connections between the situation in West Africa early in the twentieth century and what is currently taking placing in the Pacific region remains largely hidden.

The reason for this neglect regarding the imperialist character of the palm oil business in Colombia and Ecuador could be connected to some of the following facts: first, the generalized conception of how imperialism works, and the still prevailing notion that imperialism is inseparable from huge armies, large contingents of bureaucrats or imperialist agents, large-scale and intensive wars and a huge investment of capital. Second, some think that there are different moments and kinds of imperialism and that imperialism is a finished stage or was overcome by an "Empire". In this conception there are no longer states competing to expand their control and influence, but the configuration of a global hegemonic power (Hardt & Negri, 2000), or the rise of the "globalization era", in which the relations between core and periphery have been substituted by a complex multinational network through which capital, goods and people move. The third reason for the neglect of the category of imperialism is the way in which agribusinesses such as palm oil have been studied and analyzed. There is a general notion that capitalist investments in land, food and agricultural production represent a new sphere for capital accumulation, a situation stimulated principally by the peak in oil prices and the search for new energy sources. The fourth and last factor is the political context in both countries, principally in Ecuador, where for some the idea that a country governed by a leftist government may be subject to imperialist relations is simply not admissible.

Today, the situation in the region this research is about is remarkably similar to what occurred in West Africa during the establishment of the oil palm plantations. The discourse on the civilizing effects the industry brings to people and the environment is also strikingly similar, as are the racialist structure of production and the distribution of profit.

Those similarities are not just a coincidence, and of course are not an isolated case: the way in which oil palm agribusiness functions on the Pacific coast in Colombia and Ecuador corresponds to the "primary" scheme of creation and appropriation of capital accumulation within the capitalist world-system. But what are the concrete connections between the palm oil business in the Congo or Nigeria in the late nineteenth and early twentieth centuries and the current palm oil business in Tumaco and San Lorenzo? What is the connection between the palm oil business that operated in West Africa in the mid-nineteenth century and the model operating in the Pacific today? What is the relation of a production model that operated within a context of struggle between countries with one in which investors and beneficiaries are mainly local and national elites and where production is mainly aimed at supplying national or regional demand?[85]

This part of the research addresses these questions. This chapter uses and resumes the work of classical, "newly" classical and contemporary analyses of the features of the structure of imperialism today in order to explain how this structure works inside the palm oil agribusiness.

My argument is that imperialism must be understood on the one hand as a set of relations through which the unequal transfer or exchange of ecology, goods, and labor is performed. This set of relations allows the ruling class to appropriate those structures of production that support the ecological, biological, material, symbolic and ideological structures in peoples' lives. On the other hand, as a multi-geographical structure, the capitalist class is

---

85 Colombia exports just 30% of its palm oil production. The main buyers are the UK, Spain, and Germany (Mingorance, 2006). Ecuador exports more than 50% of its crude red palm oil to Colombia.

located everywhere, even being composed of local, national and regional elites, who are also from the "Third World". This means that the notion that imperialism necessarily implies a relation of subordination between countries in the Global South to countries in the Global North is debatable, as are the orthodox conceptions of the core and periphery relations.

Despite the vast literature on imperialism and unequal exchange, those analyses have rarely been accompanied by empirical studies.[86] It is necessary and possible to present the way in which imperialism works using empirical data; these days agribusiness offers a wonderful chance to do so. This empirical verification can be made, as I demonstrate, by tracking how a small village in the Pacific signs away its ecology and social resources – soil nutrients, water, life diversity, labor force, etc. – into the balance sheets of business interests. This empirical verification helps us to discover that beyond financial dependence and technological subordination, imperialism implies a relation of domination in which people-communities and nature, collective work, individual labor, as well as the environment are exploited and expropriated in order to allow for the accumulation of capital for a small group of benefactors at the core of the system.

Imperialism and the palm oil agribusiness

Rosa Luxemburg and Vladimir Lenin were two of the first authors to analyze the global dimension of capitalism. Both authors became pioneers of a world-system perspective further developed by authors such as Braudel, Arrighi, Wallerstein, Amin and others. Luxemburg emphasized the role played by a society not regulated by capitalist production (not ruled by wage-labor) in capitalist accumulation, exploring the details and effects of the competition between capitalists to gain control of the access to raw materials, cheap labor and markets in order to realize surplus value (Luxemburg, 1978). Lenin, in turn, focuses principally on the product of this competition, explaining the reasons and ways in which finance capital, and the monopolist structure over the market, takes a predominant position

---

86 Some of the few attempts include the work of Bunker (1984) on the Brazilian Amazon; Arrighi (1979) on the Peripherization of Southern Africa; and Foster and Clark (2009) on nitrate and guano in Peru.

within the system (Lenin, n.d.). While Luxemburg emphasizes the relations between non-capitalist societies with the capitalist system and the conditions through which non-capitalist societies are key to the accumulation of capital, as well as the violence this process contains, Lenin pointed out the way this process occurs inside the advanced capitalist countries in Europe and in the case of the USA. He analyzed how and why a few sectors of the ruling class (banks and trusts) advance in terms of capital concentration and monopoly practices over global production and circulation, thus establishing and benefiting the accumulation of capital.

Despite the differences between Luxemburg and Lenin, both authors agreed that imperialism is neither the product of particular policies developed by some particular state, nor is it produced by the desire of a specific sector of the capitalist class. For them, imperialism is the normal characteristic of a system that constantly needs expansion in order to increase its rate of profit; imperialism is an intrinsic part and feature of the capitalist system. That fact does not mean, however, that states or particular sectors of the capitalist class do not play any role in the development of imperialism. For both authors imperialism constitutes a kind of class agenda, in which states and enterprises work together, in order to allow the ruling class to control access to cheap labor, raw materials, commodity production and trade in markets around the world.

Imperialism is a power structure organized in a hierarchical scheme of labor and the distribution of profits. By using violence and other methods of coercion, a small group at the core of the system controls and profits from the production not only of goods, but also of bodies and nature in the rest the world. Most of us are in some way familiarized with the history and effects of European colonialism since its conquest of the so-called "New World". We know how the gold from America, together with the labor of native peoples and Africans, made the development of Western Europe and the USA as the center of industrial, economic, political, and military power possible. We have some ideas about how imperialism has been working since the early expansion of the world-system, and how through wars, dictatorships, and armed occupation, the imperialist order has been holding on and expanding. In short, we have the big dramatic stage of imperialism.

Yet, as Roberto Marin wrote, imperialist domination cannot be reduced solely to its more visible expressions such as foreign capital, political and technical subordination. Imperialism is the basis upon which the capitalist system has been functioning not only in Latin America but also in Africa, Asia and Eastern Europe (Marin & Milla, 1995; Luxemburg, 2003; Harvey, 2006). Together with exceptional situations such as military intervention, occupation, *coups d'état* and dictatorships, there is the normal operation of the capitalist system, the normal activities of businesses and markets. Since the arrival of Spanish Conquistadors, even a remote village in the Pacific coast of Colombia and Ecuador has been inserted into this scheme of production and distribution of labor and profit.

In this order of ideas, we should be asking how a bank in London profits from a small town on the other side the Atlantic; in other words, how is surplus-value extracted from a village such as Tumaco-San Lorenzo and then transformed into capital on the New York or Tokyo stock exchange? The idea of a London banker traveling in a boat through the rivers Mira and Cunapi, or Mataje, buying tonnes of palm oil to then sell them at the stock exchange in New York or Frankfurt is very tempting, and certainly this was the situation at the beginning of the twentieth century.[87] These days however, many businesses are managed by companies whose owners belong to regional elites from Colombia, Ecuador, Brazil, Mexico, Argentina, Chile and so on.

Therefore, together with capitalists coming from the USA, Western Europe, Japan, and Russia travelling around the world searching to increase their capital, there are the regional capitalists claiming their part in the division of the world – the BRICS are an example of that. Sometimes they are competing side by side, at other times they make harmonious alliances consolidated through business partnerships or marriages.[88] Regardless of citizenship the search for profit became the goal for all, making the regional identity of capital disappear, if it ever existed at all.[89]

---

[87] We probably could write a version of Joseph Conrad's novel, The Heart of Darkness for the region, relating not the Scramble for Africa and the beginning of the massive oil trade in the mid-nineteenth century, but also the *Scramble for America*.

[88] Sometimes it is hard to find the difference between each one.

[89] Recently the expression *Latinization of capital* has been coined referring to the fact that many of the richest capitalists around come from Latin America, e.g. the Colombian Luis Carlos Sarmiento, one of the biggest investors in palm oil

The world-system theory and its further developments, mainly on unequal exchange theory, have focused on understanding the historical configuration of the hierarchical structure of capitalism, especially by analyzing the role that peripheral countries have been playing in this system. Yet although the unequal exchange concept introduced us to a theoretical perspective of the relation between core and periphery within the system, its analysis does not truly allow us to see what exactly is being exchanged, and most of the time it does not offer empirical data to understand what kind of elements take part in these exchanges. In other words, beyond the perspective of labor and raw material prices, the question as to why unequal exchange constitutes a disadvantage for peripheral societies remains unresolved.

Using the example of palm oil plantations, I attempt to provide an analytical and empirical framework to understand how imperialism works through unequal exchange. I explain what is being exchanged in the palm oil business, and how this exchange functions. In other words, I show how one tonne of palm oil produced in Tumaco-San Lorenzo turns into capital, and what this transformation implies.

Imperialism, nature and the labor force

One characteristic of oily liquids is the ugly trail they leave behind, which almost everyone hates because they are hard to remove. But trails are also clues, they can be used as "breadcrumbs" to find the way back to the road. The second chapter tracks the "oil crumbs" in order to understand the expansion of the palm industry from West Africa in the late nineteenth century to Tumaco-San Lorenzo in the twentieth century. In this chapter the trail

---

in Colombia, listed in sixty-fourth position by Forbes. But individuals such as Sarmiento express only the fact that capital does recognizes neither nationalities nor geographical borders, and that social scientists love to coin new terms. Now, as Walter Mignolo (2013) illustrated during the Society for Latin America Studies conference in Manchester, capital is no longer exclusively in the hands of white people. Many of the most powerful capitalists in the world come from Latin America, Africa, and the Middle East. This situation should reorient our conception of core-periphery: What do core and periphery mean inside this new geopolitical configuration? I made some propositions on this point in the first chapter arguing that today there is a global process of peripherization whose main feature is the prevalence of unequal exchange over expanded production.

brings us to understand the imperialist relations in which palm oil is involved, scrutinizing how a tonne of palm fruit in Tumaco-San Lorenzo means capital and riches for some but destruction and poverty for most others.

This section seeks to elucidate how a small village in the rural area of Tumaco-San Lorenzo is involved in the world-system of capital accumulation. It attempts to explain what exactly the capitalist class appropriates in places where the relations of productions are not dominated by the capitalist mode of production, and where workers earn less than they need for their basic reproduction. The aim here is to bring empirical support to the analysis of imperialism and unequal exchange. Then, using empirical data, the text argues that both expanded production and primary accumulation are possible only through a recurrent and continued unequal exchange based on the exploitation and dispossession of surplus values produced under non-capitalist relations and structures of production.

Nature in the palm oil business

In the last three decades, Western societies have become increasingly aware of the impact of manufacturing production on environmental conditions on the whole planet. They have realized that this risk affects not only people in poor and peripheral countries, but also those living at the core of the system. Thus, the environment and ecology have become issues on the political stage as well as on the market. Still, economists use the euphemism "negative externalities" or "unfortunate side effects of" to denote what they call the negative and unaccounted costs of industrial production (Ruggiero, 2013). The negative effects which are hidden by this expression include climate change, photochemical ozone (smog) creation, eutrophication, depletion and acidification of soils, toxicological stress on human health and ecosystems, the depletion of resources and water sources.

In order to reach a better understanding of what the causes of these negative effects really are, and to balance and calculate their impacts, some scholars have developed the LCA (Life Cycles Assessment) methodology. LCA is a technique promoted with the stated objective of compiling, examining and finally reducing through mathematical models the "inputs" and "outputs" and the flows of materials, energy and waste in industrial production.⁹⁰ LCA is becoming very popular in some areas of production such as oil and bio-fuels because it could function as a kind of certificate of environmental awareness for corporations.

The LCA methodology is used in the palm oil industry to analyze the effects of plantations, to compile information on environmental exchanges (emissions, resource, consumptions), and to calculate and interpret indicators of the potential impacts associated with such exchanges on the natural environment (Pleanjai et al, 2007). LCA as used in the palm oil industry aims to evaluate the environmental performance of the different phases of oil production, i.e., plantation, extraction and processing. On the plantation, which this research is focusing on, the "inputs" described are fertilizers, herbicides, diesel, and in some cases water. The "outputs" defined are emissions and oil palm fruit.⁹¹

Nevertheless, the LCA methodology and focus is very problematic because it merely describes a limited range of so-called "inputs" and "outputs" – energy, water, raw materials, goods and waste gas emissions involved in industrial production. It does not allow us to see the whole panorama and refrains from questioning where the "inputs" come from and where they really go after the production process. Where does the water needed for palm oil plantations come from, and where do pesticides go? Another problem is that this methodology neglects the social relations implicated in the production, omitting for example the labor force in the description of energy resources.

More than a methodology that describes flows of matter, energy and waste in the industrial production, what is required is a format that allows for the analysis of where matter and energy originate and go, and which criticizes the distribution of benefits, profit and losses. It must also be able to discover and thoroughly expose the social and environmental implications of industrial production. Combining the study of the environmental

---

90 Further information on LCA is available at http://www.gdrc.org/uem/lca/lca-define.html
91 I am using as a basis the work of Pleanjai et al (2007); Flores (2008) and Acevedo (2012).

conditions necessary to make a plantation succeed[92] with the analysis of the social relations involved in the plantations, this section describes each step in the growing of palm, tracking the seed from the time it was planted until the time the palm bunches entered the oil mill.[93] The aim is to do a kind of inventory of visible and invisible elements in the palm oil plantation, thus exposing how the distribution and appropriation of "inputs" and "outputs" function at local-regional and global levels.

More than a description, I attempt to make visible the socio-bio-environmental relations involved in palm oil production in the region of Tumaco-San Lorenzo, exposing the effects this plantation has for people, communities and nature. By tracking these elements, we can enter the "magic room" of the agribusiness and see the way in which capital and wealth are produced on the one hand, and poverty and destruction on the other.

Soil, water, rainfall and sunlight

The production of one tonne FFB requires an area of approximately 4,000 m² of land with the above mentioned characteristics. Oil palm plantations are dependent on a complex combination of mineral, organic matter, gases, liquids, and a myriad of organisms that have been interacting and changing over long periods of time (Chestworth, 2008). The soil used on plantations requires specific levels of organic matter and also of minerals such as Na (Sodium), P (Phosphorus), K (Potassium), Mg (Magnesium) Ca (Calcium), S (Sulfur), B (Boron), Fe (Iron), Zn (Zing), Cu (Copper), Mn (Manganese) and Al (Aluminum).

The Pacific region is rich in organic matter, and in minerals such as Al (Aluminum), Ca (Calcium), K (Potassium), Na (Sodium), it has, in addition, a favorable 'cation'-exchange capacity.[94] Each tonne of FFB absorbs huge

---

92 A deep study of environmental issues would require technical information and also the professional training I lack.
93 In order to have a good delimitation of the object of research here, the work does not focus on the process of production, modification and selection of seeding. But it is important to mention that the seeding selection takes between 18–20 months, and is very complex and expensive due to genetic engineering, one of the most important issues in the palm oil agribusiness.
94 The relative ability of soils to store one particular group of nutrients, the cations, is referred to as *cation exchange capacity* or CEC (Mengel Fundamentals of Soil

amounts of nutrients existing in the soil and benefits from the interactions of organic and inorganic matter.

Huge extensions of forest and jungles have been razed, reducing the capture of CO^2 emissions. Fertilizers such as nitrogen, which is applied to the plants during the nursery and field growth phases, destroy the living conditions of other plants and animals, and generate large emissions of N^2O into the air, contributing to global warming. Other key elements that are essential for plantations are rainfall, water sources and sunlight. EG grows in warm weather, maximum temperatures ranging from 29–32.778 °C and minimum temperatures of 22–24 °C, at altitudes of less than 500 meters. Oil palms grow only in places where the rainfall is 2.000 mm or more, and there is a minimum of five hours of daily sunlight. One tonne of palm oil fruit requires 1,100–1,400 m³ of water, and it is calculated that the production of one tonne of palm oil needs approximately 3,653 m³/t. The estimation of the virtual water trade in the global palm oil business is 1,067 million m³/year.[95] Fertilizers like nitrogen and phosphate, together with other chemical products used on plantations, generate leachate. Leachate mixes with groundwater, which flows into water sources such as lakes, rivers, and estuaries, poisoning them and affecting activities such as artisanal fishing, with which many people in the region, one the wettest in the world, make a living.[96]

Once capitalists begin to use a hectare of soil to cultivate palms, whether the land was bought paying the best price on the market for it, occupied

---

Cation Exchange Capacity, https://www.extension.purdue.edu/extmedia/AY/AY-238.html).

95 Virtual water trade is defined as "the volume of freshwater used to produce a commodity, measured at the place where the product was actually produced" (for further information, see Hoekstra, 2008; for the case of the water footprint of the palm oil business, consult Arevalo, 2012).

96 The Water Science School of the U.S Geological Survey (USGS) have reported that an excess of nitrogen overstimulates the growth of aquatic plants and algae, which in turn can clog water intakes, use up dissolved oxygen as they decompose, and block light to deeper waters. The respiratory efficiency of fish and aquatic invertebrates can be affected, leading to a decrease in animal and plant diversity, and affects the use of water. In addition, too much nitrogen in drinking water can be harmful to young infants or young livestock. Excessive nitrates can result in the restriction of oxygen flow in the bloodstream (USGS Water Science School: http://water.usgs.gov/edu/).

illegally, or simply stolen, they are using for their personal and private benefit resources such as soil, rain, sun and weather. All these elements have been produced free of charge for capitalists, they exist independent of human intervention, and there are no monetary values to measure their worth. Capitalists are profiting from natural resources that, as Shiva (1991: 10) describes "are produced and reproduced through a complex network of ecological process, which remain beyond the scope of the production system for the market economy".

The lack of attention to nature's role in the industrial production system is connected to the blindness towards what kind of materials and substances are being returned to the environment, and how waste and trash are affecting life on this planet. In a market-oriented system, nature is understood as merely a commodity, a pantry of raw materials for the production process, or, at best, as a green commodity. People profiting from the exploitation of this resource do not worry about what will happen to the waste and residues; they will leave as soon as the party – and resources – are over.

Palm oil industrial agriculture in the Tumaco-San Lorenzo region, as in many other places, is putting wild species in danger, exhausting water sources, and depleting soil, generating at the same time the proliferation of plagues, deforestation, and also huge amounts of liquid and solid waste. This process is global in nature, but its effects are particularly dramatic in peripheral places, where in general, capitalists assume no monetary responsibility for their impacts.

The labor force: Individual and collective production

Once the area of cultivation is selected, soils must be prepared to plant seedling palms. These plants have spent between 10 and 12 months in the nursery being irrigated, fertilized, cleaned and selected; women mostly carry out these activities. Men, who must remove weeds and plow and level the field, mostly carry out soil preparation. After cultivation and during the time the palms are producing healthy fruits, 12-15 years, plants and soils need schematic and rigorous control to ward off weeds and plagues. Every week leaves must be cut, fallen and damaged fruits collected, and schedules relating to pollination, fertilization and the control of diseases have to be maintained.⁹⁷ The complete description of this process requires some technical terms which exceed the possibilities and aims of this research, but it is important to mention that the maintenance of soil and plants is done using technical tools many of which small farmers have no access to due to their high costs, and which in some cases can be dangerous to their health.⁹⁸

Palm starts bearing bunches 2½-3 years after being planted, and the bunch must be harvested every three weeks. In general, harvesting tasks are carried out by teams formed by three people; one cuts off bunches using a Malay knife, while the second worker carries the bunch to be transported by animals and the third worker collects fallen fruits. Those activities constitute the "unskilled labor".

On large plantations, people doing harvesting work earn, at best, a monthly salary not exceeding 200 USD. On small plantations, sometimes farmers themselves do this task, with the help of relatives, and sometimes by workers the farmers hire on a daily basis. The income for small farmers producing palm oil is between 500 and 800 USD monthly.

Around the world, labor conditions in the palm oil industry are frequently and strongly criticized, even being compared with slavery. There is a wide body of documentation on the poor working conditions, low pay, and disadvantages that small producers have vis-à-vis the large companies (Canas, 2009; Pye & Bhattacharya, 2013; Rainforest Action Network et al, 2015). All of this criticism is important, but it is problematic: usually the criticism comes from a liberal, but not a critical perspective, which does not challenge the idea that labor is a commodity to be sold. According to this view, the goal of society should be to get the best price possible for it. They simply appeal for better salaries and working conditions, more rights for the workers and a better distribution of the profits, as well as more investment

---

97 The effects of these activities on workers' health were described above.
98 One of the chemical products used in palm oil plantations is glyphosate. The molecule glyphosate was patented by Monsanto in the early 1970s as the active ingredient in the herbicide Roundup. Today, together with the growing use of glyphosate in agriculture, the doubts and questions as to the danger this product poses for humans, animals and soils is also increasing, as is the debate on the political and monetary power that Monsanto is creating, and how this company thrives by controlling and monopolizing seeds, soils, and other natural resources (Filomeno, 2014; Robin, 2008).

in infrastructure, schools, hospitals and so on. In a way this discourse naturalizes and sanctifies the wage-work relation. Yet, just focusing on the wage-labor structure as the core of economic relations means neglecting the entire structure that makes the existence of "workers" as biological entities and social beings possible.

People working for palm oil companies, either as hired labor or as "small independent producers", are not just a labor force, producers of commodities and earners of a salary. These people are human beings resulting from a complex network of social-bio-ecological structures. A person nursing the plants, cutting and collecting palm oil bunches has a body, this body was given birth to by a woman, nursed, fed and loved by her and by a group.[99] Most of the time the body is born capable of assimilating, processing and reacting to information the world provides it; it is a body that is able to learn. The group the body was born into offers it conditions to strengthen its capacities, and to shape the body's behavior; the body becomes skilled, and those skills allow it not only to understand the world, but also to manipulate and change it.

Monthly incomes earned by workers and farmers, (in this case $250 for hired workers and $800 for farmers) do not cover basic or even minimum conditions, what some would call a living wage,[100] to reproduce energy spent during the production process, and to support a family's needs: food, housing, medical care, clothes, education, childcare, transport, recreation, and a retirement pension.[101] Neither do they cover additional needs for the development of a social group such as infrastructure and access to art and culture. How do people fill the gap between what they earn as wage-labor-farmers and what they and their families need? How do they meet the rest of their needs which are not covered by their salary. The testimony of one small farmer provides us with some clues:

> Casi todo lo que desayunamos hoy (chocolate, plátanos fritos, queso, huevos, maíz), lo producimos acá en la finca. Bueno, solo falta el aceite con el que fritamos,

---

99   Women also almost always do these activities.
100 To find out more about living wages see Anker, 2006.
101 In the previous chapter, I exposed that the region Tumaco-San Lorenzo has one of the highest poverty rates and NBI (Necesidades Básicas Insatisfechas) in Latin America.

ese lo compramos. Yo cultivo palma y saco tiempito para otros cultivos, también tengo animales ... Acá nos ayudamos entre todos; mi esposa, mi mamá, hasta mis hijos cuando llegan de la escuela me ayudan en la finca (Farmer in Tumaco. Interview. Januar, 2013).

This farmer's words show that together with the production of palm oil, they carry out independent activities to benefit from natural resources. Fishing and hunting with farming work on their own plots or as hired labor provides workers-farmers and their families food and basic housing. Together with activities related to providing food, there are networks of communitarian and family-based work which provide childcare, care in old age, recreation,[102] and so forth. The gaps between the basic needs of the labor force are to some extent covered by the worker himself; both the workers and small farmers have to work for themselves in order to fulfill the necessary conditions to work for the palm industry, meaning they are *working to be able to work*. But along with the work done by direct workers, there is the work done by families and the community that guarantees *the worker the basic conditions to keep working*. All of this direct, indirect, personal, collective and communitarian work constitutes a whole universe of social production that is further transformed into private property for the capitalist.[103]

---

102 Antonela Picchio (2003) makes an approximation of the incorporation of unpaid work in the analyses of the economic system. She understands as unpaid work "the upkeep of living spaces and domestic goods, care for health, education, and the psychological needs of family members, and the maintenance of social relations". She uses a simple co-operative circular flow to show a relationship of interdependence between families and firms. In this circular flow relations of exchange are shown both as monetary and of another nature; the firms buy labor and sell goods, the families provide labor and buy goods. It is assumed that this flow is produced on the basis of co-operation between the two institutions for the reciprocal interdependence of their interests. But she does not speak about accumulation. There is no reciprocity between families and firms. Where does the surplus come from and go to? This is a blind spot in Picchio's analysis.

103 Nevertheless, the work done by workers themselves and by their families and communities just cover a limited number of people's needs. Other needs, such as formal education, medical care, drinking water and a sewage system, roads, and garbage management, are hardly covered due to the infrastructure requirements which depend on government institutions. This lack of basic services and infrastructure is part of the blockage I mentioned previously.

Behind one worker doing an "unskilled task" on the large plantations or producing fruits to sell to the companies, there are complex and invaluable networks of biological and social relations of production. These relations of production not only bring to life a nursed, skilled and socialized labor force, but also keep it alive. In a context where income and salary do not cover basic needs for biological and social reproduction, we have structures of production that generate the workers' conditions in which to continue working. Those are relations that support the "workers" whether they suffer the collateral effects of working for the palm business or not, such as, poisoning, illness, injuries, and when the person laboring is no longer of use to the companies.

By analyzing in detail the connection between social production and production of palm oil as a commodity, we can see capitalists appropriate surplus produced at three levels: first they use a nursed, skilled and socialized living labor force, one ready to work even though the capitalists have not paid a dime to guarantee basic living conditions. The extreme poverty people live in there makes this blatantly obvious. Then, during the working day, the labor force produces amounts of exchange commodities, surplus value which becomes the private propriety of the capitalist. And finally, by paying less than the labor force would need to reproduce itself to recover the energy used in the production process, capitalists are extracting additional benefits from those elements that supply the gaps between worker-farmers' income and what they need for basic subsistence.

Palm oil fruits produced in large plantations or by small farmers are possible because this production combines the appropriation and over-exploitation of natural resources with the appropriation and exploitation of social relations that bring to life a nursed, skilled and socialized labor force, as well as the overexploitation of living work. One tonne of palm oil, which turns into private stock for the company, contains countless socio-biological structures which need a lifetime to exist and develop, and which have been produced without any charge to us, being in most cases a result of collective and communitarian activities.

It means that although there are some public schools in the region, they are far away from the communities, and the quality of the education they offer is constantly questioned.

It is said that palm oil is a legal and legitimate business in which "free enterprises" hire "free workers" and "free farmers" in a frame of liberty and equality. But, as has been exposed, the palm oil agribusiness is shaped by exploitation, dispossession and violence. The figure of free commercial relations, expressed by a signed or verbal contract, in fact hides the entire environmental, biological and social elements capitalists appropriate and for which they do not pay. Farmers and workers sign contracts with enterprises, but those contracts do not mention the complex universe of socio-bio-ecological relations, which are embodied in the labor force, nor the violence used by capitalists in order to appropriate them. The idea of free choice also hides the whole political, cultural and legal structures erected around the agribusiness to legitimate it. Thus, the relations between the Black-Afro workers-farmers and the whole community and palm oil companies is an unequal exchange because it becomes a "zero-sum game" in which all loss of utility for worker-farmers, the community and the environment constitute the gain of palm companies.

The magic of capital

In Chapter IV of the second section of the first book of Capital (1887), Karl Marx analyses how money is transformed into capital, studying the character of labor-power as a commodity and the way in which, in the capitalist mode of production, labor-power produces surplus value that is transformed into capital. Marx understands labor power as "those mental and physical capabilities existing in a human being, which he exercises whenever he produces a use value of any description" (Marx, 1887: 119). But, those "mental and physical capabilities" Marx refers to exist thanks, as we have previously pointed out, to a social-biological structure of production, and not as a consequence of this human being's actions *per se*. One needs only to imagine, for example, that a newborn is abandoned, left uncared for: without any question the infant would die some hours or, at best, a couple of days later, depending on the situation. This human being was not capable of producing the basic elements needed for his or her survival. Those "capabilities" Marx refers to are not something intrinsic to human beings, but the product of socialization and also of collective and communitarian work; Marx (1887: 120) himself was in some way conscious of that issue when he wrote:

> Nobody – not even "a musician of the future" can live upon future products, or upon use values in an unfinished state; and since the first moment of his appearance on the world's stage, man always has been, and must still be a consumer, both before and while he is producing ... In order to modify the human organism, so that it may acquire skill and handiness in a given branch of industry, and become labor – power of a special kind, a special education or training is requisite, and this, on its part, costs an equivalent in commodities of a greater or less amount. This amount varies according to the more or less complicated character of the labor-power. The expenses of this education (excessively small in the case of ordinary labor-power), enter pro tanto into the total value spent in its production.

My proposal is, following on from Marx's understanding of labor-power as a source of surplus value, to go a step further by exploring in greater detail the whole structures that make the existence of labor-power possible and to then bring these insights into the analysis of capital accumulation.

As was shown in the first chapter, one of the features of capitalism as a world-system is the articulation of different regimes of control over nature and labor to produce surplus. Around the world there are many particular ways of appropriating natural resources, social relations, and overexploitation of living work either directly or indirectly. The most unashamed forms of slavery and servitude may be combined with community and family work, and with formal and modern wage-labor relations. At the same time the most modern technique for exploiting nature coexist alongside the most rudimentary ones.[104]

Labor and nature control regimes are designed and administered by a variety of groups, who guarantee the flow of socio-bio-ecological elements to the core of the system. They are responsible for distributing profits and losses, and, from different positions and functions, guarantee the continuity of capital accumulation. Thus, despite having different positions and incentives, the whole universe of agents from small farmers, policymakers, company owners to a whole spectrum of workers, managers, scientists, bureaucrats, and even "unskilled workers" take part and support the chain through which nature and labor are extracted to become capital and wealth.

---

104 The preponderance of one kind over the other depends on the position or place in the world-system.

A capitalist investing in palm oil receives much more than palm oil fruits as returns. On the one hand, they receive the surplus value produced by workers during the working day as Marx described, on the other, all of the socio-bio-ecological elements used during the entire production process which exceed the normal 'working day' on plantations. But, in the same sense that workers are not just workers, those elements such as soil, water, sunlight and others required for palm oil plantations are not sterile and dirty pieces of waste, nor exchange commodities produced by some company. These elements constitute the basis for life on Earth, and are a product of a complex network of interactions that requires millions of years to come into existence. They also exist independently and free of charge, though this is not without its limits.

How has it been possible that natural resources which all living beings, not least humankind, depend on, were transformed into a private pantry and the property of a small collection of individuals who do not use it to supply any living need but rather commercialize it and make profits? How has it been possible that private agents profit from the whole network of social relations that foster life, nurse it, teach it skills and socialize it as labor power? How do people allow themselves to be overexploited? How do external agents not only appropriate the direct produce of their work, but also the produce of communitarian labor at the expense of generating a latent risk of destruction of the socio-environmental structures?

Capital is a social relation. This relation has a material and an immaterial component. Its material side is the sum of all goods and commodities that are properties of the capitalist and which he will commodify in order to increase his amount of capital. These commodities are the result of combining the appropriation of direct and indirect labor and nature.[105] The immaterial

---

105 Control over nature and its overexploitation has been a major factor since the early phases of the development of capitalism. The relevance of this fact for Marx and Engels is clear in their work on the development of capitalist agriculture and its effects in Europe as well as in the rest of the world, as can be seen in the following paragraphs: "All progress in capitalistic agriculture is a progress in the art, not only of robbing the laborer, but of robbing the soil; all progress in increasing the fertility of the soil for a given time, is a progress towards ruining the lasting sources of that fertility". (Marx, 1887: 330) "As long as the individual manufacturer or merchant sells a manufactured or

component of capital, which precedes and overtakes its material expression, creates and recreates the conditions which make possible that a particular group of individuals can appropriate, accumulate and profit from the whole socio-bio-ecological production. As a social relation, capital is intent on potentially subordinating the whole of production, distribution and consumption, converting it into monetary relations and making the search for profit the absolute measure of all things. The immaterial form of capital shapes social relations in which everything must prove its productive viability, or perish if it fails to do so (Mészáros, 1995).

Within the capitalist system, the main goal of social relations is production for the market in order to attain the maximum profit, using money as a common denominator. By combining ideologies with violent coercion, money has established itself as *the* regulator of social life, and the increase of capital and the unlimited search for profit has been declared as the core of social relations. These social relations created by capital have in turn been mystified, exposed as something natural, ahistorical and perennial, something superior to human forces, hidden under a shadow of uncontrollability (Mészáros, 1995). The capitalist class, with the help of scientists, academics and policy makers, have established their own and particular interest as the general interest of the whole of society.

In the palm oil business, scientists and the academic community play a key role in the ideological arena. In the name of progress, partisan scientists,

---

purchased commodity with the usual coveted profit, he is satisfied and does not concern himself with what afterwards becomes of the commodity and its purchasers. The same thing applies to the natural effects of the same actions. What did the Spanish planters in Cuba care, who burned down forests on the slopes of the mountains and obtained from the ashes sufficient fertilizer for one generation of very profitable coffee trees – what did they cared that the heavy tropical rainfall afterwards washed away the unprotected upper stratum of the soil, leaving only the bare rock!". Dispossession of farmers and native communities of their means of life, that is, land, water, soil, nutrients, was the way to guarantee the availability of labor power at the time of colonialism and imperialism and are the acute expression of the struggle between Western European countries and the USA for control over raw materials. Thus, the current commodification of water, seeds and the environment are unfortunately merely the continuation of a process that started long ago and pertains to the inner logic of capitalist accumulation.

as Shiva called them,[106] established profit maximization as the "rational" end of human action, promoting and legitimating the exploitation of humans and nature.[107] The work of the scientific community has been supported by policy makers, who develop legal frameworks through bureaucratic officials, and through the building of the necessary infrastructure to make those possible.

At the front line of palm oil expansion have been policy makers, scientists and research institutes. They cooperate, which means they influence not only palm oil production itself, but also the spectrum of the control and dependence without which it would not be viable. They train people, collect information and design technologies to exploit people and nature faster and cheaper. Then they monopolize these tools to sell them to the best bidder.

For its part, the government apparatus (law, bureaucracy and military) establishes the general interest as the particular interests of the capitalist class. This corresponds to the current impulse to monetize all aspects of life, justified through the work of the scientific community in disqualifying the use-value-based disposal of labor and nature that communities have. Bureaucrats, policy makers and scientists work together in most cases to label communitarian non-profit-oriented production as irrational, non-modern, unsustainable and contrary to "the best interests of the nation and to general progress".[108]

---

106 With the concept "partisan science" Vandana Shiva (1991) refers to the sciences, whose goals and methodology are guided by monetary interests, a science for which natural resources are insulated and the non-interacting nature acquire economic value only when commercially exploited.

107 In the article "Technology and science as ideology" Habermas (1987) exposes Marcuse's analysis of the formal concept of "rationality" (presupposed aims in given situations) as the expression of political domination, and the role played by science and technology as ideology and ideological tools used to control, dominate and destroy nature and society in favor of a ruling group.

108 Of course, within the governmental apparatus as well as in the scientific community there are counter-hegemonic positions and actions. There are scientists, most of them social scientists who criticize the capitalist system, and give legitimacy to non-profit-oriented ways of life, systems of organization and production. Nevertheless, they are a marginal group within the whole scientific spectrum.

This has been the framework that configures palm oil plantations as the ultimate receipt of most socio-biological relations of production in Tumaco-San Lorenzo. But, as I have been pointing out throughout this text, this agribusiness is just the most recent expression of a phenomenon that has been surfacing for centuries. The Pacific region has for a long time been a pantry for the exploitation of nature and labor power within the capitalist system. Here, foreign and local capitalists have worked together to spoliate communitarian, collective and non-profit-oriented relations of production to generate capitalist accumulation and profit. Developing an argument to explain why this process is repeated again and again constitutes the main aim of the next chapter.

Conclusions

The argument I have developed in this section can be summarized by the notion that imperialism is not a political project of particular nations or states, but is rather an intrinsic feature of the normal operation of the capitalist system, characterized by constant patterns of unequal exchange. This is the framework under which the palm oil business works in the tropical coastal regions of the Pacific. In these regions people "freely accept" taking part in this exchange because physical and ideological violence is constantly employed against them in the form of forced displacement, threats and murder, as well as by means of the institutional disqualification of non-profitable production relations.

The way imperialism is portrayed today, i.e., with a fixed notion of where the core and the periphery are located, is contrasted by the fact that core-and-periphery relations are currently being expanded globally at the same time as a peripherization of production relations is taking place, issues that demand further research. In this section, I showed how unequal exchange is carried out by revealing all the socio-bio-ecological elements involved in the production of a particular commodity, in this case, palm oil.

My argument here was that capital has two aspects, one material (sum of goods and commodities), which is the result of the combination of the expropriation of nature, appropriation of socio-biological elements that support the labor force, and the direct and indirect exploitation of living work. The material entity of capital creates and recreates those conditions that allow a particular group to appropriate, accumulate and profit from the whole of socio-bio-ecological production.

By analyzing and placing center stage the relation between the capitalist system, those structures that make possible the existence of the means of production (nature) and the labor force, the parasitic nature of capitalism itself becomes clear. Exposing the way in which all socio-bio-ecological production becomes private property weakens the ideological basis that gives the system legitimacy. Contrary to the liberal and neoliberal perspectives, those structures that constitute the source of capitalist accumulation exist without any cost to us whatsoever are not structurally profit-oriented and are the result of collective work.

Chapter IV. Capital Accumulation and Socio-Ecological Resilience

> *"Según he sabido, desde que nos arrancaron de África nos vienen correteando; mi bisabuela y mi abuelo me contaban todo lo que les tocó huyendo de la muerte, y nosotros en las mismas. Sinceramente ya estamos cansado de correr, qué es lo que quieren de nosotros."*[109]

At this point, the way in which the book is progressing might seem to be contradictory to the reader. While the first chapter was dedicated to "dismantling" the predominance of "capitalo-centrism", reclaiming the need to see the complexity of social production, the following two chapters dealt mainly with the palm oil industry as part of the capitalist system; the global, transnational and regional history of this business; and its development in the rural territory of Tumaco-San Lorenzo. The way in which the two previous chapters were developed corresponded to my methodological and analytical aims. In Chapters Two and Three I attempted to delimit, as much as possible, my research aims. I tried to exhibit as much information as possible to understand the history, functions, structure of palm oil plantations, the particularities of this crop as a commodity, and its implications for people and nature, particularly in the region the research is about. This process was long and exhausting, but key to obtaining empirical elements to answer the main question this research is about: the conditions that make the continuity of the "primary schema of capital accumulation" possible.

Some elements important to understanding the continuation of the process of capital accumulation in its "primary or primitive" form were addressed in the first chapter. Supported by the work of Luxemburg and Gibson-Graham, I attempted to break down some postulations used in socio-economic analysis about how the accumulation of capital works. I reintroduced, as Luxemburg and Gibson-Graham did, the role that non-capitalist structures play in the capitalist accumulation system. I showed

---

109 Testimony by Teresa, a woman who lives in Tumaco and tells her family's story in the region (see the complete testimony in Burgos, 2010: 204).

that, despite the difference between their work, these authors have common conceptualizations regarding labor power that overcome classical, neoclassical and even Marxist critical perspectives. For both authors the production of surplus value for the market under the wage-labor form is just one part of complex networks of production, circulation and consumption.[110]

The fact that production of surplus value to accumulate capital under the wage-labor form constitutes just one level of social relations of production sheds light on how socio-economic production continues to be a narrowly explored field and that many key questions about social reproduction still remain unanswered. In that sense, my question was, if the production of goods through the wage-labor relation to generate surplus value constitutes just a part of the whole social production, where do those elements that allow the reproduction of human beings as biological and social entities come from? In other words, if the main goal of capitalism as a system is to generate profit and not to guarantee optimal conditions for the reproduction of human beings (nor labor-power), where do the conditions that support our existence come from, and how and by whom are they provided?

In the third chapter, on the basis of the palm oil industry, the research provided a concrete example of where the social-biological production is going. The research exposed how complex nets and relations of social-eco-biological production are used to produce a specific commodity. I then exposed how soil, water, climate, and skilled and socialized bodies interact with each other to elaborate a good, in this case a tonne of palm oil fruit. Furthermore, I argued that in oil palm plantations, the accumulation of capital occurs behind the usual stages used to analyze the production of surplus value. Those elements, constituent of capital, do not appear on the scene during the observation of how plantations work. Inside the palm oil plantation, it is possible to see a particular chain of energy exchange, on the one hand, input of energy (labor) and on the other hand the output of energy (commodity), but that scenario does not allow us to see the other relations of production – those provided by labor-power and nature.

---

110 Of course the way in which both authors made the emphasis is different. While Luxemburg's goal is to integrate non-capitalist production into the analysis of capital accumulation, Gibson-Graham's is to expose that the majority of those structures do not work in a market-oriented economy.

The latter could mean that the scenario generally used to analyze the production of wares and commodities has been hidden in the whole universe of production involved not only in the production for the market, but also in the production of human societies.

In oil plantations, either those owned by big companies or small farmers, we see a labor force producing a particular commodity, but the whole picture is hidden. We do not see where the labor force came from, where those living, socialized, skilled bodies able to work are coming from. We know that workers earn less than they need to reproduce the energy spent during the labor process, but by analyzing the production relations in the orthodox way, it is not possible to see what the workers do to close the gap between what they need to live and what they earn. Thus, how the reproduction of labor-power is possible in a context of overexploitation, where workers earn the minimum to continue living, remains a mystery. On the other hand, the analyses of a commodity as just a social product neglect those elements produced by nature such as soil, water, and sunlight, crucial needs for a successful plantation.

The palm oil agribusiness provided an opportunity to analyze how different modes of production are interacting to produce surplus value and profit. By exploring as exhaustively as possible the operation of palm oil plantations, the research collected enough empirical and theoretical data to argue that in the palm oil agribusiness the main resources for capital accumulation are derived from non-capitalist modes of production. This means that all the elements shaping the palm oil fruit as a commodity are mainly a result of non-capitalist relations of productions, which are collective and not oriented to searching for and maximizing profit. These structures and relations of production not only existed previous to the scenario of simple and expanded reproduction described by Marx and Engels, but a *condicio sine qua non* capital accumulation could not be performed. Therefore, capitalism is a system in which the whole socio-bio-ecological production is transformed into capital, then accumulated and appropriated by a particular group.

This chapter, the final one, aims to introduce some issues that hardly appear in the theoretical analysis of production relations, the production of social human beings and the production of nature. Both processes constitute a fundamental task in social science because, as economic political feminism claims, the analysis of capitalism as a world-system has omitted essential elements without which the capitalist system cannot exist: i) the reproductive capacity of the ecosystem to supply natural resources and ii) the reproductive capacity of society to supply the human workers (Dunaway, 2001; Picchio, 2003).

This chapter is divided into three parts. The first one presents some ideas I want to discuss related to the conditions that make the continuity of primary accumulation possible. I also put forward some concepts such as social metabolism and resilience to analyze this process. The second part introduces some basic elements to understanding the process of the production of the environment and nature. The third part deals with the production of bodies and Black-Afro society. Here, the research shows the relationship between exploitation for capitalist accumulation and resilience, and analyzes how the Black-African population have displayed all their capacity for resilience to confront and overcome recurrent cycles of exploitation and dispossession.

Socio-ecological resilience and capitalist accumulation

To produce a tonne of oil palm fruits and attain profit from it, capitalism creates particular environments and relations between human beings and nature. The nature (or: environment) of Tumaco-San Lorenzo has been altered, regulated and disciplined. Something similar has occurred with labor-power: now most of them are, as the former boss from FEDEPALMA wished, subordinated to an alienated working day as worker-farmers. The mono-plantation has alienated labor power from the means of production, nature, and the final product. Oil palm plantations express that the capacity capitalism has to configure "kinds of ecological system" as Harvey (2014: 247) states:

> Capital is a working and evolving ecological system within which both nature and capital are constantly being produced and reproduced. This is the right way to think of it. The only interesting questions then are: what kind of ecological system is capital, how is it evolving and why might it be crisis-prone?

I agree with Harvey's idea, if he understands an ecological system as a specific form of organization of the relation between people and nature that makes the obtention of capitalist profit possible. Harvey forgets that capital cannot produce the ecological elements or the labor-power needed to configure the "ecological system", and actually, without the capacity of nature and the labor-power to reproduce themselves outside of the profit-oriented relationships, businesses like palm oil would not be profitable. To produce a commodity, in this case oil palm fruit, it is necessary that people and nature exist. That healthy, socialized and skilled labor-power and the environmental conditions have been "produced", as well as the production of those balanced environmental conditions that allow the existence of the planet as we know it, and the successful repetition and continuance of those processes.

These days, despite the expanded control that humans have over the/their environment and the control that capital is gaining over seeds, fertilizers, pesticides, bio-technologies, the processes and relations that make life possible still overcome human capacity. Despite the commodification of many aspects of life, the aspiration of capital to make everything marketable and profitable still remains just a wish, one, it is worth saying, which is constantly challenged worldwide. Despite the fact that laboratories and pharmaceutical industries have a high capacity to alter seed features and to manipulate the environment where the plants grow, the success of a seed depends on the natural chain producing environmental elements such as soil, moisture, temperatures, precipitations, etc. In short, palm oil depends on bio-ecological production that make the flow of matter and energy, and on social production, which configure the labor-power capacity to manipulate this flow possible.

With these complex production relations as a framework, I now attempt to provide possible answers to the topic this book is about: the conditions making possible the repetition of the primary accumulation process of capital in the Tumaco-San Lorenzo region, as well as its particularities and implications today.

Capitalism is mainly a system of accumulation that is in need of constant expansion in order to find cheap labor, raw materials, and new markets to maximize a given rate of profit. The capitalist system is characterized by an imperialist hierarchical scheme that operates under many labor regimes and modes of production around the world. This system is based upon an unequal distribution of profit and loss, as well as of benefits and wastage. Core positions profit from the system's social-bio-ecological production existing mainly in those places and societies localized in peripheral positions.

Between the core and the periphery, there occurs a constant unequal exchange, or an unequal flow of energy and matter, because those societies localized in peripheral positions not only have to reproduce their own living conditions, but also to exceed that which constitutes the fountain of capital and wealth for the core.

From the early expansion of the capitalist world-system to this day, large areas of Latin America have been subjected to the global accumulation system through an unequal exchange. This exchange implies a permanent extraction of matter and energy, raw materials and a labor force without any equivalent compensation. I argue that the continuity of this exchange is possible because this matter and energy still exist, seeing as people and nature are in a condition of being reproduced outside of the logic of capital. I argue that nature and societies have some capacities that help them cope with external stresses and disturbances. Thanks to these capabilities in the region of Tumaco-San Lorenzo, nature, groups and communities can, after "traumatic" events, reproduce and maintain their living conditions. Thus, the continuation of the production of environmental conditions, together with the abilities of Black-African groups to produce socialized bodies in a fit condition to work, have provided, from colonial times to the present day, the excess that during every boom-bust cycle is inserted into the global capitalist system.

Production of nature[111]

In general, human beings believe themselves to be the core, the alpha and omega of life. But our existence is a very minuscule part of the large, complex and changing planet. Planet Earth, its nature and its environment have been here for a long time, longer than we have. None of us was involved in the development of the conditions that make life possible here, and probably, despite the chaos we create, the planet will exist long after we disappear.

Without pretending to be an expert on environmental issues, but with the conviction that social science research needs a holistic focus, and to break with the anthropocentric-functionalist perspective, in the next

---

111 This section was written using as its main reference the work of Acosta (1959); Alcaldía de Tumaco (2008); López, et al (2008).

section I describe some nature-environmental elements in the region this research is about. I introduce basic but key elements to understanding some basic facts about the landscape on the Pacific coast as well. This section invites us to think about the complex production structures which have been shaping this region for millions of years and allowed the accumulation of capital for over five centuries. It also attempts to take into account processes generally overlooked in socio-economic studies or dealt with through a reductionist perspective in which nature and the environment only exist for the purpose of some human beings' wishes or needs; food, tools, clothing, housing and medicine.

The tropical region of the Pacific, from Panama to Peru is a zone that has hardly been studied, documented and researched. This region, as was stated in the second chapter, has been marginalized by the political and economic elite since the colonial occupation a situation that has prevailed despite the many commercial trade booms. Political isolation has also contributed to its marginalization in scholarly research, and despite growing interest in the last two decades, the zone remains a "mystery" for social and natural sciences as well as for businessmen and policy makers.

The few studies that exist on the natural and environmental features of the region have been executed by private businesses; many of them oil companies, which barely allow any access to the information they collect. On the other hand, much of the available literature on bio-ecological topics has been available for more than three or four decades. Many ideas and interpretations have still not been updated by new visions and methodologies. In this way, much of the good research about the region has become obsolete or anachronistic. Another problem is the nationalist focus the majority of those studies have; the political-administrative division has meant a partitioned understanding of places sharing similar natural bio-ecological features. Because of this situation, one has to spend more time and energy in order to be rigorous and provide an articulate point of view of the region; I myself have had to search for the same information twice, once for Tumaco, then for San Lorenzo.[112]

---

112 Some scholars such as West, and in some cases Friedmann and Whitten tried to break with this tendency, analyzing both sides as a whole. But the trend is to view the region as being split in two.

It has been difficult for me, a social scientist not skilled in Earth or natural sciences, to attempt to understand and communicate issues related to geology, biology, ecology, etc. Nevertheless, overcoming my own limitations, I have attempted to understand the region the research is about as a whole. Of course it is nothing new to say that there is a need to talk about nature and the environment to understand production relations, indeed, the majority of research carried out on the region contains a chapter, subchapter or paragraph describing nature-ecological and environmental aspects. But many of them still see nature and the environment as an appendix to human beings, not valuable in themselves.

There are some theories that postulate that the Earth is a complex, self-regulating system, which maintains surface conditions favorable for contemporary life. As a system, the Earth has the capacity to keeps its climate and chemical composition in a relatively stable internal state (Lovelock, 2009). There are many debates and confrontations concerning the scientific plausibility of these theories, but I find some of them very useful because they provide us with a holistic – not an anthropocentric – point of view about networks, structures and relations of production of life.[113] These theories not only challenge the traditional criterion that confines the biosphere to a passive role. They also assign non-living organisms active roles in influencing the global environment, and introduce the notions of cooperation, mutuality and symbiosis as definitive characteristics and forces in evolution, questioning Darwin's idea of nature as "red in tooth and claw" (Bjornerud, 1997).

From this perspective, it makes sense to claim that to understand the way in which capital accumulation has been performed for centuries, we must incorporate those processes that make the existence of natural elements such as soil, water, sunlight possible, as well as the complex interaction between organic and inorganic matter. The key point about the valuation of nature is to understand how natural resources are being extracted to be integrated into the capitalist world market (Boyer, 2015). Those environmental resources used to produce commodities, palm oil in this case,

---

113 Gaia theory was born in the early seventies by James Lovelock and others. Basically the theory claims that the planet Earth is a system possessing conditions to regulate itself in order to maintain surface conditions favorable for life.

in order to make profit are the output of millions of years of interactions between living and non-living beings, and are also the result of mutual, symbiotic and dynamic non-profit-oriented relations of production.

Geomorphology, climatology and biota

The Tumaco-San Lorenzo region is part of the Tumaco-Borbon drainage basin. A basin is an area in which surface waters converge at a particular and single point on a lower elevation. The drainage basin topography is a product of the interaction between precipitations and solar energy. This landscape is determined by the interaction of both precipitation and solar energy as well as by the resistance from the topographical surface. These elements also define the basin's altitude, the constituent rocks it has, as well as the percentage of plant coverage and presence of a layer of soil.

Beyond narrow geological information, some researchers calculate this region was formed during the late Cenozoic, in the Paleocene epoch (approximately 26 million years ago) as a result of the convergence of the Nazca and South America plates. These conclusions are based upon the study of rocks, sediments and deposits from different geological formations including Paleozoic, Cretaceous, Paleogene, Neogene found in the area (Suárez-Rodriguez, 2007; Lopez et al, 2008). The current geomorphological aspect of this region is a product of the alluvial sediments deposited during the last million years.

Tumaco-San Lorenzo is comprised of three geomorphological zones: littoral, alluvium, and hills-mountains. The littoral zone (0 meters AMSL) is constituted by a dense mangrove swamp mainly in the north, and an arid coast in the south, which contrast with the rich variety of fish offered by the Pacific Ocean. This landscape is continuously being shaped by tectonic movements, the rain of volcanic ash, climatic changes, and human settlement (Usselmann, 2010). The alluvium zones are places between the inland areas and rivers where soil or sand is deposited by running water formed during the Holocene. These zones constitute rich deposits of valuable ores such as gold and platinum. The hilly landscape is formed by small mountains between 100 and 400 meters above sea level (Corsetti, 1990; Alcaldía Tumaco, 2008).

Tumaco-San Lorenzo is a region affected mainly by two meteorological phenomena; El Niño and La Niña. Both are currents of warm or cold ocean water that develop in the central and equatorial Pacific. While El Niño generates high humidity levels and intense rains, La Niña produces periods of drought. In this region the temperatures oscillate between 22–32 °C and precipitations levels are between 2,313–2,792 and 3,000 mm, constituting one the wettest places in the world. This combination of factors makes the region a very humid area with levels humidity in the air of between 84% and 88%.

Regarding its bio-geological structure, Tumaco-San Lorenzo is part of Tumbes Choco-Darien. It is a tropical rainforest set between two large rivers the Mira and the Mataje, which is a complex chain of small rivers, brooks and lagoons, such as Mejicano, Gualajo, Rosario, Caunapí, Bogota and Santiago flow from. The majority of these rivers are navigable and constitute the main means of transport for local people. Some experts argue that the vegetation in this region represents one the highest biological developments on Earth, being part of one of the largest active centers of speciation and endemism around the world, estimated at a minimum of 8000 species of plants (Barnes, 1993; Gentry, 1989).

Another main feature of the tropical forest in the lowlands of the Pacific coast is that it has a high capacity for self-reconstruction, which as Acosta (1959: 50) explains and claims has not yet been deeply studied:

> The forest land is used for agriculture and if abandoned then shapes a stover. … If the stover remains abandoned for many years or definitively, it is slowly covered by native plants that finally eliminate the exotic ones, and in this way the forest naturally aims to restore its primary composition.[114]

According to Holdridge's "life zones classification"[115] there are three main formations or bioclimatic schemes in this region: tropical rainforest, tropical wet forest, and moist forest. Furthermore, the tropical rainforest (bh-T from 0 to 200 Meters AMSL) covers approximately 45% of the region and is localized near the coast. Here, high average temperatures and a significant

---

114 My own translation.
115 Holdridge's concept of life zones is a global bioclimatic schema for the classification of land areas. It is becoming one of the most significant schemata to demonstrate the key importance of climate in evolution, selection and plant growth (see Kircher, 2013).

amount of rainfall characterize the ecosystem. The tropical rainforest in Tumaco-San Lorenzo in particular has high levels of biodiversity: a single hectare may contain up to 42,000 different species of insects, 313 species of trees and 1,500 species of plants (Phillips et al, 2007).[116] The vegetation in this area includes more than seven types of mangrove forest and peaty forested wetland, locally called Guandal forest. This zone is formed by three kinds of forest, Sájales, Cuangariales y Guandal, but there is also the presence of subsistence agriculture crops such as cocao, coconut, plantains, as well as grassland for livestock and palm oil plantations.

The mangrove area constitutes a particular and unique type of forest due to the confluence of fresh and salt water. Mangrove tree species have adapted to these conditions, and are dependent on salt water to exist. In the Tumaco-San Lorenzo region, the species *Rhizophora mangle* is predominant, but other species such as *Laguncularia recemosa* also exist, and the less common *Pelliciera* genus which has only been seen in the Pacific region of Costa Rica, Colombia and Ecuador. The mangrove forest provides refuge and food to many birds and mammals. Due to the density of the trees' roots which prevent access to big predators, it constitutes a kind of nursery for fish, shrimps and arthropods. These forests are also very important to the cycle of rotting organic material because they help to retain many contaminating elements and shield the coast from hurricanes and cyclones which are very frequent in the region (IIAP, 2012).

The tropical wet forest (bmh-T) is a large range, which crosses both villages parallel to the coast. In Tumaco this forest covers an extension of 168,000 hectares, equivalent to 50.1% of the area. The vegetation distribution in this area is similar to that of the tropical forest; nevertheless, it is less dense, and exhibits smaller trees. Forest, subsistence agriculture, palm oil plantations and grass for livestock also cover this area. Moist forest (bp-T), the smallest one, is an area localized close to the mountain range. This so-called eco-region has a high influx of precipitation; annual rainfall rates fluctuate with higher averages in the north, while the area becomes progressively drier in the south. The area is influenced by intense volcanic activity, and trees with a dense canopy, rich in lianas and epiphytes, many

---

116 Acosta made one of the most extended classifications of trees.

of which are endemic, dominate the vegetation. In this kind of forest, 1,250 plant species have been recorded as corresponding to 136 families, and of these species 43 are endemic (Dodson & Gentry, 1993).

Fauna

Similar to the geomorphology and vegetation, the fauna in the region of Tumaco-San Lorenzo is a narrowly studied field. The small amount of research conducted in this area is limited to those species which could be of interest for commercial trade. There are few facts known about the wildlife here. As part of the Tumbes Choco-Darien, the region of Tumaco-San Lorenzo has one of the most species-rich areas on the planet. It has an exceptional abundance and endemic fauna including birds, arthropods, amphibians and insects.

As was said, this text does not intend to do a detailed report on the large list of species and families of animals present in this area, it simply attempts to highlight its diversity and complexity. This region is home to more than 127 species of amphibians mainly belonging to the family *Bufonidae, Hylidae* and *Le Ptodactylidae;* almost 100 species of reptile, of which the following species stand out: *Chelydra sp, Kinosternon sp, Lepidochelys olivácea, Caretta caretta, Dermochelys coriácea, Eretmochelys imbricata, Iguana iguana, Basilicus sp, Ano lisantonii, Boa constrictor, Bothorops asper, and Caiman crocodylus.* There also more than 577 species of birds living there.[117] Further, the habitats of more than 186 freshwater species, 66 of them endemic are to be found in the area (Maldonado et al, 2012), as is the case with saltwater fish such as crustaceans and mollusks, which mainly live in the mangrove forest. Lastly, the region features a rich species of mammals, many of whom have been barely studied and are in danger of

---

117 According to research carried out by the Instituto de Investigaciones Ambientales del Pacífico in 2012, some of the most representative species are the *P. olivaceus, P. occidentalis, Penelope spp, Amazona spp, Pionus menstrus, Phalacrocórax olivaceus, Quiscalus mexicanus, Amazilia tzacatl, the Psittacidae, the Scolopacidae, Charadriidae, the Phalacrocoraz olivaceus, Podilymbus podiceps, Casmerodiusalbus, Egrettathulay bubulcos ibis, Sporophila spp, Nyctassa violácea, Elanoides foficatus, Aramides sp, Charadrius spp, Columba spp, Geotrygon sp, Pionus menstrus, Amazona spp, Amazilia spp, Ceryle torquata, Progne sp, Quiscalus mexicanus and the Ramphocelus icteronotus.*

extinction. They could disappear before we know they have even existed. Some examples are tamarins, tapirs, anteaters, monkeys, pumas, ocelots, leopards, and jaguars.

As one can see, beyond being just a pantry of raw materials for the global trade, the region of Tumaco-San Lorenzo is part of an organized, balanced and equilibrated community of life. Here, after millions of years, life has attained an exceedingly great degree of diversity and a high complexity of interrelations (Holdridge, n.d.). Behind every gram of gold or platinum extracted here, behind every tonne of fish, ivory nuts, lumber, timber, cocoa, bananas and palm oil, there is a large structure of bio-ecological production which works in a coordinated and harmonious fashion to generate and maintain the chemical, atmospheric and climatic conditions of existence for animals, plants and minerals, as well as for human beings.

Palm oil business entrepreneurs profit from these production structures that generate conditions that make every type of life on planet Earth possible. The soil where the palm is planted and the nutrients it receives, the water used for irrigation, the sunlight, the ability the palm has to interchange matter and energy, all those processes are part of a system with self-organizing and adaptive behavior. The combination of heat, light, moisture and soil on the earth, together with the interaction between organic and inorganic elements make the existence and the continued reproduction of environmental features everywhere possible.

Insofar as those conditions continue to exist, it is possible that this region will continue to produce either raw materials for the market-oriented system or a means of life for the communities living there. Without the coordinated, collective and harmonious work of the planet and its biota, there would be no chemical, atmospheric and climatic conditions that make the existence of fauna, vegetation and minerals possible here. Without a system with self-organizing and adaptive behavior with a high capacity to process matter and energy and the ability to absorb disturbances, neither the basic scheme of production, circulation and consumption nor the global market-oriented system would be possible.

This issue could be seen as self-evident, but unfortunately it is not. Up to now we have rarely understood that every human social formation, from remote communities of collectors, hunters and fishers to the complex, global capitalist exchange system is dependent on the bio-ecological conditions the Earth provides. We are also dependent on the ability of resilience the ecological system has which allows it to essentially retain the same function, structure, regeneration, and therefore identity while it experiences shocks and disturbances (Walker et al, 2006).

To conclude, due to its bio-geological features, the Tumaco-San Lorenzo region produces raw materials and food for trade on the market. Take the distribution of the ivory nut trees, for example: the fast reproduction and particularities of this hard big nut made it possible to convert this region into a provider of the global ivory nut trade (Leal, 2005). The capacity the forest has to restore its composition by recovering native plants and trees has made the development of the lumber and timber business possible. Its capacity to process matter and waste has made this forest an ideal place for the development of large plantations of banana, cocoa and oil palm.

Production of society

I started the text with the promise of explaining the history of destruction and damage, but also the reconstruction and resilience of the Black-Afro communities living on the Pacific coast shared by Colombia and Ecuador in the rural areas of San Lorenzo and Tumaco. Up to this point, I have only dealt with one part. The research has described how since the colonial period, this region has been a treasure-trove of raw materials and cheap labor for the global capitalist system of accumulation. Also, I have shown that palm oil represents a new cycle of colonization and exploitation inserted into the imperialist scheme of power. The text has also analyzed how by using violence, political and monetary domination, the capitalist class profits from the entire socio-bio-ecological production structure, and how nature and labor are transformed into capital and wealth for a few, and poverty and destruction for many others.

This section is concerned with the second part of the process. It describes and analyzes those tools and strategies used by the Black-African population to adapt themselves to a hostile environment, but also to confront and resist the effects generated by the "cycle of extractive activities".

In their work on timber exploitation in the Pacific region, Leal and Restrepo (2003) argue that the extractive economies – mining, ivory nut, lumber, etc. – the economies for accumulating capital have depended on practices, relations and visions that the Black-Afro communities have developed over these territories.

> La extraccion de recursos naturales para el mercado extraregional y las ganancias asociadas, sólo ha sido posible en la medida en la que se han apoyado en el trabajo de las poblaciones locales, es decir en sus prácticas, relaciones y percepciones del entorno (Leal & Restrepo, 2003 131).

Similar conclusions are reached by Taussig (1979: 40) in his studies on the structure of the sugar cane plantation in the department of Valle del Cauca, to the north of Tumaco, in Colombia.

> [L]a agro-empresa ha dependido de la formación de una clase de campesinos pobres que combianan la mano de obra asalariada en las fincas capitalistas, con su propia producción campesina en pequena escala ... los semiproletarios y productores campesinos le están suministrando también un subsidio a los hacendados capitalsitas, y es este subsidio el que permite que los hacendados capitalistas puedan obtener ratas más altas de plusvalía de lo que sería posible si los costos de sostenimiento y reproducción de la mano de obra fueran completamente dependientes del modo de producción capitalista.

This work aims to go beyond Leal and Restrepo and Taussig's work by understanding the links between extractive-accumulation and subsistence economy. I argue that the market-and-profit-oriented system could only establish profitable business for more than four centuries up to the present day due to the ability the people in this region have developed to continue producing a social structure based on a non-capitalist vision of nature. This implies that the processes of production of social beings and also the capability these people have to hold a material and cultural production, allow not only the subsistence and continuance of the groups, but also generate conditions that have made the continuance of primary accumulation possible.

For some readers, the latter argument could be problematic due to some of the argumentation the text dealt with in the first part. In the first part of this research, following the work of authors such as Gibson-Graham, Federici, Picchio, and Dunaway, I argued that social organization and the reproduction of the laboring population are based on a complex network of non-capitalist structures. These structures are not profit-oriented, a fact which makes individual and social reproduction possible. In this context, the ability to continue producing healthy, socialized and skilled bodies, as well as a symbolic and material structure based on a non-capitalist vision of society and nature, is not a particularity of the region studied. If, as I argued, primary accumulation understood as the appropriation of values produced by non-proletarianized labor under non-capitalist relations of production is the basis of capital accumulation, that means that primary accumulation is happening constantly around the whole world, and not only in the geographical periphery of the system.

Related to the first issue, I want to point out that the main aim of this research is to understand the conditions that make the continuity of the primary form of accumulation possible in the world-system. The case of the palm oil industry in Tumaco-San Lorenzo provided an exceptional opportunity to explore in-depth how the accumulation of capital works and to show how capital accumulation is based on the expropriation of socio-bio-ecological production. Palm oil serves as an example to show that without the reproductive capacity of the ecosystem to supply natural resources and the reproductive capacity of women and communities[118] to supply human laborers and consumers, the capitalist system could neither exist nor be profitable.

In addition, I remarked in the first chapter that we are witnessing a global process of peripherization characterized by the prevalence of unequal exchange over expanded production. In that way, the region studied constitutes a window to understanding a global process and its effects on society and nature, but also the forms through which society and nature confront it. Finally, nowadays the capitalist class is global in origin, the owners of palm plantations belong to national and regional elites involved in many other kinds of business, who share positions of power and interests with the elite of Europe, the United States, Japan, and so on. This fact shows that the "core-periphery" relation is something determined less by geographical features than by one's position in the chain of production and distribution. In that respect, to hold a peripheral position within the system implies having an obligation to use the socio-bio-ecological production

---

118 The reproductive capacity is a feature of some women, but not all of them, and giving birth is just one part of a group of elements needed to form a human being. Caring for and socializing a human body is a collective task done mainly by people who did not give birth to the children; grandmothers, aunts, relatives, and kindergarten and primary school teachers.

structure to provide surplus to the core, receiving less than necessary to reproduce one's labor.

From the arrival of European conquerors to the present day, the region of Tumaco-San Lorenzo has been in a peripheral position in the system. Here, nature and the labor power of the black population filled the coffers of the conquerors with those resources demanded by international trade, i.e. gold, platinum, ivory nut, timber, lumber, banana, cocoa, fish, palm oil, etc. But the population have also sought out living conditions for the members of the community during and after every extractive boom.

I introduce the concepts *social metabolism, resilience* and *metabolic rift* as a framework, in the next section, the last of this project, in an attempt to analyze how the Black-Afro population has survived, confronted and challenged the continuing process of exploiting social production from the slave trade to the present day.

Social metabolism and resilience

The word metabolism (Stoffwechsel) appears in Marx's work (Capital: 192); it denotes different but connected issues. Sometimes it refers to the cycles of the vegetation's exchange of nutrients, it can also be a description of the interaction between man and nature, where a labor process is a process of exchange of energy and materials; and at other times is an analogy to describe the circulation of commodities under capitalism.

Marx's ideas about the relationship between nature and society, about the process of exchange between human beings and the environment, or better still, social metabolism, were taken up by Georg Lukács (2000) in the twenties and István Mészáros' in the seventies (1995), and today by John Bellamy Foster and Joan Martinez-Alier.[119] One of the core ideas behind the concept

---

119 Both authors are concerned with disassembling a generalized idea of the separation between ecology and Marxism and expose clear connections between political ecology and historical materialism, tearing down Marx's idea of a utilitarian anthropocentric, a blind belief in the development of productive forces at any cost. Bellamy Foster and Martinez analyze how the work of Marx and Engels was highly influenced by the work of botanists, agricultural chemists and physicists such as Sergei Podolinsky and August von Liebig, as well as the debates on deforestation and the degradation of soils in Europe and America, seeing as large estate agriculture was integrated into the historical

of social metabolism is that human beings are part of a continuing process of matter and energy exchange. Nature provides us with the environmental and biological conditions for existence; we consume and transform that matter, and finally return it as waste. Then this matter is processed by the environment, thus facilitating the continuity of the cycle. Nevertheless, unlike other living beings, the exchange between humans and nature features the ability we have to transform some conditions through labor. This means we not only receive but also adapt the biological environmental conditions that make our reproduction as individuals, communities and species possible, but at the same time transform nature and ourselves (Marx, 2004; Engels, 1883).

The concept social metabolism refers to those interactions between human beings, society and nature that make human life possible. These interactions are modified by the human being through the development of production forces. We, as a species, are in a position of deciding what we consume, as well as how to control what we return to the environment. We are also in a position to orientate our labor force beyond reproductive activities, creating particular types of cultural and ecological experience. Hence, depending on the technical resources certain groups have, as well as the prevailing rationality, the exchange of matter and energy can be used to guarantee the continuance of social metabolism or to disturb and even destroy it.

On the other hand, the concept of resilience has been defined as the ability of ecosystems, groups and communities to cope with external stresses, disturbances and changes. This aspect is gaining importance as a component in the analysis of the circumstances under which nature, individuals and social groups adapt to change. But the term also concerns how in "adverse", uncontrolled and disturbed situations, ecological and social systems attempt to hold and rebuild conditions that allow the individuals' and the groups' continuing existence.

Originally the concept was used in ecology, but recently it has been extended to social research, particularly on how human beings react to "catastrophes and natural disasters". Today, there is a growing field of research which analyzes the connection between social and ecological resilience, as

---

materialist conception developed by the writers of Capital (Martinez-Alier, 2004; Bellamy Foster, 2000).

well as the desire to overcome what some authors call the absurd and artificial split between ecology and society (Neil, 2000; Walker et al, 2006). Despite the potential these concepts have within social science, in general, a short-term vision on social resilience predominates. Most papers and texts consulted on this topic focused on short-term experience, as well as on specific "traumatic" or "unexpected" events, such as particular wars, armed conflicts or "natural catastrophes". However, this concept could also be used within a long-term perspective, analyzing, for example, how the Black-African population in the Pacific have being dealing for almost four centuries with the constant extraction of individual and communitarian production, as well as with the violence this process implies.

This research exposes and explains the structures developed by people living in the rural territories of Tumaco-San Lorenzo to guarantee the reproduction of social and material life before, during and after every extractive "boom". Here, I analyze the conditions and strategies created by the Black-African population to generate and control the exchange of energy and materials in order to guarantee social reproduction. Then, I highlight those structures and tools developed by people living in the rain and mangrove forests to confront the constant and recurrent cycles of exploitation threatening their survival.

Cycles of spoliation, social resilience and metabolic rift

Humans, like every other organic being, need to satisfy intrinsic biological requirements to continue living. But unlike animals, plants, and other organic beings, we have developed and increased our capacity for control, modification, manipulation and regulation of the flows of matter and energy. Thus, we create particular needs for living which, although they have become more complex and sophisticated over time, one could say have remained unaltered for thousands of years. The supply of food and drink, provision of a place to store, to sleep, to repose and to meet other people (relatives, friends, etc.), the care of children, elderly, and the sick, training, the acquisition of skills and the socialization of individuals, and waste disposal, as well as the burial of the remains of dead humans, are all needs that every human collective has been successfully met for a long time, despite social change, wars, environmental crisis, chaos and destruction that constantly occurs.

This scenario of chaos, destruction, but also creation and reconstruction has in the descendants of the African people one of the most palpable examples, as Roberto Burgos Cantor (2010: 11) wrote:

> En la situatión de la población afrodescendiente, la vida vivida, la vida negada, la vida consquistada por la lucha adquieren una importancia específica, sin cuya consideración no se entiende nada. Esa vida se ha opuesto a los estereotipos, a cuantas maneras, a veces con inconsciencia, adopta la discriminación, y las trampas de la lástima...Por ello los ojos y las preguntas inquieren por una historia capaz de contar cuándo habíamos llegado aquí, y en qué medida este aquí corresponde a una ganada pertenencia...

Inspired by the lucidity with which Burgos understood the history of the black population in Latin America, this section goes deeper into the strategies and tools used by those people to continue living within a system whose starting point is constant and continued spoliation.

The Black-Afro population has not only been despoiled by the conditions that supply their basic human needs, but have also faced constant obstruction to meeting this need. The ability of the population in this region to adapt, confront, resist and counter the effects generated by the "cycle of booms and fluctuations" that characterize extractive economies has attracted the attention of some social researchers. Whitten and Friedmann (1974), for example use the concept "ethnical adaptation" to highlight how Blacks have contended with the vicissitudes that the market economy imposes on them by developing adaptive capacities and particular structures, social organization and production, such as the *Ramaje, Tronco, Minga,* and *Mano Cambiada*.[120] For the authors, although these capacities and structures were established during colonial times, they have been expanded and generalized to this day, constituting the basis of the distribution of the means of production and organization (Whitten & Friedmann, 1974).

Rivera (1986), for his part, argues that the basis of the social and economic organization in Black-Afro communities in Ecuador has been the ability of the people to maintain some independence and autonomy in the production

---

120 Explanation of what these are and how these structures work is provided below.

of goods, thereby reducing the negative effects of the capitalist markets. Escobar (2008) and Taussig (1979) have studied the ability the communities have developed to guarantee their survival by combining their subsistence production with production for the market economy, analyzing at the same time the contradictions, problems and potentialities of this situation.

The work of these authors shows that the struggle for this region did not begin with neoliberalism, or the expansion of the social and armed conflict in Colombia, nor even the rise of the palm oil agribusiness. The lowlands of the Pacific have not been the "haven" of peace, tranquility and harmony that many inhabitants, activists and academics think.[121] On the contrary, they have been an arena of historical confrontations and conflicts, which have changed in intensity, forms and dynamics. In this context, the Black-African population has settled and become the dominant population in this region.[122] They have developed their abilities and strategies to confront, resist and counter the effects generated by the "cycle of booms and fluctuations" that characterize extractive economies.

This part of the research deals with the study of the process of the continuous creation and resistance mounted by enslaved Africans and their descendants in the region of Tumaco-San Lorenzo. The analysis is divided into cycles, and shows the continuity of the capacity for resilience that the population has demonstrated, for more than four centuries.

First cycle: Slavery, survival and freedom

"El Tapao" is a dish that is part of the culinary tradition of Black-Africans in the lowlands of the Pacific coast. This dish is a mix of wild meat, green plantains and some plants such as coriander. It is cooked in an enclosed oven, and during the cooking process it must be covered with plantains leaves to avoid the leakage of steam. People say that this dish was created by Africans who were trying to escape from enslavement during colonial times.

---

121 These impressions could be seen in the work of Arocha, Restrepo, and Escobar, as well as in the narratives of many activists and leaders, provided by the PCN organization Proceso de Comunidades Negras), for example, and people living in the region.
122 Rueda makes a detailed exposition of the cycles of struggle and confrontation between the three groups (see Rueda, 2001).

To avoid being discovered and captured while cooking, they opened holes in the earth to cook and cover the food with leaves.

Today, there is still disagreement on the number of Africans kidnapped from West and Sub-Saharan African and sold as slave labor in America. The calculations oscillate between four to 100 million.[123] One quarter of the slaves, between one and 25 million, depending on the estimation, died during the voyage. Many were victims of illnesses acquired during the journey or victims of the deplorable conditions in which they were transported. Others committed suicide. Murder and infanticide were acts of insubordination, rebellion and freedom (Romero, 1995).

The majority of this mass of people came from the Senegambia region, particularly from the former Kingdom of Kongo, the Gulf of Guinea, called the Gold Coast. They were Mandingas, Yolofos, Fulos, Branes, Balants, Biafaras, Casangas, Congolee, Angolans, Arara, Carabalí, Fantis, Ashantis, amongst many others. Those populations, together with the women in Europe and the native population in the "New World", were the first witnesses and victims of the early global expansion of the capitalist system. In this context, death and enslavement were the destiny for many Africans, while survival and freedom were the main goals. They had to develop "early creative responses", as Rivera calls them, to face the conditions generated by the slave system. These "creative responses" allowed them to supply material and basic needs for survival; build familiar and social structures in the midst of a strange, hostile and adverse context; materialize the wish to attain or maintain the liberty they achieved and at the same time to gain legal recognition of their status as free people; and confront their status as criminals and fugitives, persecuted by colonial powers.

The region of Tumaco-San Lorenzo, as in every place where the slave trade was carried out, experienced riots and disturbances started by those who were striving for freedom (Whitten & Friedmann, 1974; Burgos, 2010). It is not clear when exactly, but somewhere between 1533 and

---

123 There is considerable controversy surrounding the number of Africans subjected to the slave trade. The calculation ranges from nine to 100 million. Friedmann and Arocha (1986: 33), for example, calculate a variation between three and 25 million. Germán Colmenares (1973), for his part, argues that there were around nine million.

1550 a Negro slave ship crashed on the coast of Esmeraldas. Seventeen male and six female slaves, property of the Spaniard Alfonso Illescas, took advantage of the chaos ensuing from the wreck and absconded, reaching the mangrove forest in the region of Esmeraldas. Headed by a man called Anton and then by the "Negro Illescas", the "group" dominated and made alliances with the native population this area. This union gave birth to the so-called "Zambo Republic", and constituted not only one of the first areas under the dominance of Black-Africans in the "New World", but also one of the first areas recognized as being autonomous by a colonial power (Zuluaga & Romero, 2007; Burgos, 2010; Tardieu, 2006).

Africans had to survive in a strange environment: they were in need of food, fresh water, clothing and household. This first stage of resilience implied the capacity to identify and to learn what could be eaten, how and where food and water could be found, and generate the material and social conditions for the birth of new members of the group. Anton, Illescas, and the other marooned settlers explored areas of the forest and mangroves where native groups lived. They learned from the Indian population how to distinguish between edible, poisonous and dangerous plants and animals, how to farm and to fish, and how to build houses and ships. Marooned people adopted the subsistence organization of the natives. Then they were in a position to not only cultivate plantains and maize, but also to develop abilities to hunt, fish and gather. Those activities constituted the basis of their livelihood, and probably, as has been documented in other areas of the Pacific, the *platanar* and the *chacra*[124] were the core of familial production (Burgos, 2013; Rueda, 2001).

Parallel to the necessity to feed and clothe the population, and to have access to clean water, the marooned people needed to reproduce and build a community. They also needed to defend themselves and preserve the freedom they secured and protect themselves against the aggressions of native residents. After demonstrating their military superiority, thanks to the weapons they had, Illeasca's people established an alliance and kinship with the Niguas, who helped them to defend themselves against enemy groups (Zuluaga & Romero, 2007; Rueda, 2001). African men "married"

---

124 The *platanar* is a small extension of land near the house where green bananas are cultivated.

native women and impregnated them, sometimes through consent, other times by force (Rueda, 2001).

The marooned people instructed their new allies and relatives on defensive techniques that improved their protective strategies, and organized a powerful and feared army, which despite many attempts by the Spaniards, could not be defeated. Colonial administrators were afraid of possible alliances between the Zambos and pirates, and were aware of the territorial advantages the Zambos had. For this reason, they offered Illescas the title of Governor of the Province of Esmeraldas, and amnesty for him and his relatives (Burgos, 2010; Tardieu, 2006; Rueda, 2001). Illescas rejected this offer because it did not include a general pardon for the whole population, nor did it include the recognition of freedom for all of the people under his mandate. In the late sixteenth century, the Zambo Republic witnessed the victory of its diplomacy when it attained the legal recognition of freedom and the status of free vassals. For its part, the Zambo Republic recognized the figure of the King as the highest authority, accepted the Catholic religion and swore allegiance to the institutions of the Crown (Rueda, 2001; Beatty-Medina, 2009).

Yet the feat achieved by Anton, Illeascas and others was not unique. During the whole colonial occupation there were many riots, uprisings and escapes. Almost one decade before, or after, there are disagreements as to the date,[125] other groups of Africans from Nicaragua, arrived on the Esmeraldas' coast, having escaped enslavement in conditions similar to those of Illeascas' people. But the exciting adventures of Anton, Illescas, and the others constituted only a small part of the whole history of Black-African settlement in the lowlands of the Pacific coast.

Although the task of exploring the region and transporting the enslaved African labor force to the lowlands of the Pacific began in the first decade of the sixteenth century, it was not until the late seventeenth century that Spaniards had a political and commercial interest in this area. From the late seventeenth century to the early eighteenth century the first large displacement of Black-Africans to the Pacific started and the first large wave of settlement began, mainly in remote areas near rivers and estuaries. The second

---

125 Rueda argues that the arrival of these groups occurred in the late fifteenth century, while for Burgos this event occurred in the early fifteenth century.

and definite wave was the consequence of the manumission laws in the mid-nineteenth century. The first wave of displacement was partly motivated by the expulsion of the Society of Jesus (Jesuits). In 1776, the Society of Jesus was expelled from the Spanish Empire and its colonies and proprieties were placed in the hands of (the Spanish) colonial administration. It is worth noting that in Valle de Chota Ecuador, the Society of Jesus possessed entire families of slaves. Some took advantage of the expulsion of this congregation and organized collective escapes (Charvet, 2010; Fernandez, 2001).

Another event that motived Black-African settling in the remote forest zone was the construction of the "Malbucho" roadway. This roadway connected the highlands in Ecuador with the Pacific coast, dynamizing the commercial relationship between Quito and its provinces with Panama, Nueva España and Lima. The adaptation of the land for this road implied a loss of territorial control for the Zambos, who were forced to move to remoter areas (Rueda, 2001).

But the main reason behind the resettlement of Black-African communities was the expansion of the mining frontier, which occurred within what one could call the mining circuit of the lowlands between the villages of Buenaventura, Popayan, Barbacoas, Nuquí, and Santiago. Mining exploration and exploitation generated the arrival of not only large contingents of slaves, but also free Black-Afro labor. The slave labor force was organized in *cuadrillas*, groups of people responsible for identifying and exploiting new mines. Although the *cuadrillas* were designed as a labor control scheme, over time, they would constitute a "strategy of resilience" to guarantee the material and social reproduction of the black population, as well as their existence and configuration as a collective and a community. Not only was the exploitation of gold for the mine owner organized around the *cuadrillas,* but also the material provisioning for the reproduction of the enslaved labor force and development of a group identity.

Due to the high cost and the delay in the arrival of provisions, the task of the *cuadrilla* was extended to the labor of reproducing the group. The task of the group combined mining activities with farming and fishing. The *cuadrillas* had to develop all of their creative potential to face the new environmental context to generate profits for mine owners, but also to obtain their own sustenance. From their arrival as a slave labor force, Black-African people in the region of Tumaco-San Lorenzo had to provide their own living conditions in order to become an exploitable labor force, or, as stated earlier, *they were working to work*:

> El primer momento de formación de la cuadrilla de esclavos implicaba que fuera abastecida para su alimentación con provisiones del interior, dado que su dedicación inicial debía ser en actividades mineras exclusivamente. Sin embargo, para llegar a la costa esas raciones había de pasar no menos de ocho días ... Al acabarse las raciones de carne y maiz el grupo de la cuadrilla tenía que procurarse de los recursos que daba la selva espesa del Pacífico para complementar y suplir su alimentación (Romero, 1995: 50).

Probably much of the knowledge and information Africans acquired from their original societies about fishing and farming were useful in the Pacific. Fishing for example was a common activity in many populations on the African continent. However, it was different here. Due to the features of the landscape and the context they needed to learn and create specific fishing techniques. The rivers near the *cuadrilla*'s settlement were polluted by the waste of mining, and those not contaminated were far away, implying long displacements that disrupted the working day. Some solutions Black-African found were to create fish traps, which were installed in the remote areas at night and collected the next day so as not to disrupt the work in the mines (Romero, 1995). This fishing practice is still used in the communities today.

Together with the need for food, there was the need to have a place to sleep and to store provisions. The *cuadrilla* was conceived as a labor grouping, but with time social relations emerged as well as a gender based division of labor. The chain of command in the *cuadrilla* was as follows: a master or overlord; *señor de cuadrilla*, a colonial mine-owner, generally residing in large towns outside the mining district; a white or mulatto overseer or *administrador de minas*, who in the absence of the overlord was the most important person in the town; and the *Negro capitan de cuadrilla*; a fellow enslaved person, generally a man, who was the immediate boss of the work crew, in charge of disciplining his gang, the distribution of food, and the collection of the weekly take of gold for the administrator (West, 1956; Romero, 1995).

In the beginning, the *cuadrillas* were comprised exclusively of men, but with time, due to the "emerging needs" of the groups, women were introduced. Women worked in mining, farming and in the household. Within the *cuadrilla* structure, they had a double role: they were producers of direct surplus value working as miners, and as householders they guaranteed the recovery of the energy spent by workers and provided conditions (food) to keep people working in the mines. Women were responsible for the administration of the products collected by the enslaved during their additional working day. They cooked and took care of the home as well as of the small farms where products such as plantains, maize, manioc, tubers and spices were cultivated. They also served as a reference or leader of the household and family, not only in terms of material conditions, but also in terms biological and social reproduction, generating links between individuals, and in many cases keeping the group united and cohesive.[126]

Together with the development of the ability to exploit natural resources, the region provided sustenance, the use of currency and engagement in trade as buyers and sellers constituted another key element of it. People kidnapped from Africa and their descendants in the "New World" were mainly commodities, interchangeable goods; nevertheless, they were not always passive victims of the human trade and exchange market. When it was possible, they used money-gold not only to buy goods and commodities, but also to buy their own and their relatives' freedom. It has been documented that many mine owners allowed Black-African slaves to work for themselves on the weekend, and on holy and religious days. In that way, in this region as in the whole Pacific littoral, the Black-Afro population developed a buying power which stimulated trading ventures between the Pacific lowland and the growing pre-industrial colonial cities (Whitten, 1974).

Aside from *cuadrillas*, in those mines abandoned by the enslaved, there appeared the *libertos* (or *libres*), Black-African women and men who, either by escaping or purchase, gained their freedom. Once the *cuadrilla* left a mine, because it was considered exhausted, or due to the low returns it generated, the *libertos* exploited it. They did *mazamorreo* or *barequeo* (panning), a craft technique that consisted of collecting and washing the rivers

---

[126] The work of Paloma Fernandez Rasines (Fernandez 2001) analyzes the figure of the *redibitoria* through which enslaved women by appealing to the indissoluble relations such as marriage or maternal love, requested the return of a sold "piece" (a child or husband). Nevertheless, despite the chilling information about how this figure operated in the region, the fact that the figure existed and that Black-African women used it brings new elements into the analysis of Black-Afro resistance strategies.

sediments searching for small gold pieces. Sometimes *libertos* could exploit the mine as long as they paid the owner a tax, other times they could save enough gold to collectively buy a mine (Romero, 1995).

The *libertos* were in many instances former miners, and in general the head of a *liberto* group was the former captain of a *cuadrilla*. In other instances, the *libertos* were Maroons who came from far away, who escaped from their owners, and to whom the environmental features of the region as well as the existence of freed Black-African communities would provide some protection and security and a possible social structure to be integrated into (Zuluaga & Romero, 2007; Burgos, 2013).

The existence of *libertos* generated constant tensions for the dominant social structure. They had close relations and contact with the enslaved labor force, many of them were married couples, lovers, or relatives, or were involved in commercial exchange and cultural activities. This proximity disturbed the slaveholders, who considered *libertos* a bad influence on the enslaved crew (Romero, 1995). Beyond the discomfort the *libertos* and the enslaved dancing and drinking alcohol together caused amongst whites, one could think that the tension had deeper motivations. The presence of an autonomous Black-Afro population mining, trading, hunting and fishing for themselves not only motived the enslaved to escape, they also exposed and expressed the weakness of the slave system, laying bare a gap in the racial-colonial power relations.

By escaping, the *libertos* showed the slaveholder's lack of power, and by using the exchange system which had been developed by colonialists to buy their freedom they showed that to be enslaved was neither a natural condition nor divine will, but a status that could be overcome. Probably because of this, the figure of *libertos* has been a pillar for the formation of personal and collective identity. To this day this figure plays an important role in the historical narrative of the communities and constitutes a counter hegemonic figure, challenging the idea of enslavement as the foundational basis of towns, villages and cities (Escobar, 2008; Charvet, 2010; Garcia, 2010).

The survival, settlement and consolidation of freed groups of Black-Africans were possible by combining resistance with adaptive and creative strategies. They learned from the Indians how to farm and fish, they used the labor organization structure developed by the colonizer to create their own social cultural networks, and their insertion into the trade market as autonomous buyers and sellers provided them with the ability to exploit money. Social structure, as well as the organization of production established by *libertos* constituted the basis of shelter and colonization for the Black-Afro population in the wet littoral. Many familial and communitarian organizational systems operating today, as well as the exploitation of the environment and the distribution of land and natural resources, have been based on those strategies first created by the *cuadrillas* of slaves and then by *libertos* for more than four centuries ago.

Today, despite the changes and transformations generated by internal and external factors, familial networks and rights to settle and to exploit certain areas are based on *Troncos* and *Ramajes*. These are groups of people who are linked to a common ancestor, woman or man, living or dead, but not necessarily blood relatives, who were the first owners of the land and mines (Whitten & Friedmann, 1974; Romero, 2000). From the Tronco-Ramaje structure emerges the figure of *"Familia extensa"*, extended family, in which kinship is not necessarily defined by biological motherhood or fatherhood, but by a complex network of social relations built around proximity and cohabitation,[127] such as the *mina*. The mina is a living and working area created by *libertos* and slaves, and still constitutes a socio-productive unity shaped by the hamlet the miners inhabited. The *mina* is complemented by the *chagra*, the place where crops are planted and managed, mainly by women. Many working days are still organized in a communitarian fashion and operate under the figure of *minga* and *mano cambiada*, an exchange of tasks and collective work.

Second cycle: Manumission, freedom to die and the free market

The trip between the rural area of Tumaco and San Lorenzo is like time travel. Modern buildings and infrastructure where palm oil companies and

---

127 During my stay in San Lorenzo, I noted the relevance these relationships still have. After trips to school, groups of children met in the street to play. They stayed outside until late into the night, and spent the night together in those houses where they played the last game. The host family takes care of accommodation and food. Before going to sleep the children said "good night auntie" to me, to which I answered "good night nephews and nieces". This was the first time we had met, we were not actually relatives.

research institutions operate coexist with the ruins of the old depressed infrastructure built for the storage and transport of exports such as tagua, banana, plantains, etc. Next to the modern roadways, there are simple and homely dwellings, hut constructions. Many of them, despite being extended and "improved" with concrete, still conserve the legacy of the first Black-Afro settlers.

From the late eighteenth to the mid-nineteenth century, in the middle of the confrontation between white-mestizo groups fighting for independence and the Spanish Crown administrators, the promise of freedom for Black-Africans was a strategy used by both sides to recruit Black-Africans as soldiers and helpers. Many enslaved people, mainly men, took part in the conflict and went to the battlefield. Unfortunately for many of them, the time to enjoy freedom was as long as the battle itself.

The colonial administration was defeated, and in its place emerged a fragmented creole-mestizo elite for whom the promise of freedom for the Black-African population was not only not a priority, but also a matter of divisive and internal conflicts (Tovar, 1994). Black-Africans had to wait more than 40 years (1810–1851) for the legal, formal and complete abolishment of slavery. Abolition was in fact a result of a structural crisis of the system. The large slave traders were put out of business in the late eighteenth century, and the prohibition of the slave trade was extended to the colonies, which made business more expensive and risky. In addition, riots and the mass escapes of slaves were more and more frequent. The old productive system was becoming obsolete in comparison to more flexible and efficient forms of labor control (Leal, 2005; Colmenares, 1979).

Certainly, manumission implies abolishment of slavery as the core of social production, but it did not transform the social structure or the conditions in which Black-Africans lived. On the contrary, it exacerbated them. Many of the effects can be seen today in the statistics on poverty and inequality. The link between the enslaved person and the slaveholder was broken, but the socio-economic conditions did not change. Those women, men, and children who produced wealth for the elite did not get any compensation or basic tools to use to survive. After being used to generate endless capital for the core of the system and its agents in the colonies, after being systematically expropriated at all levels, Black-Africans were thrown into the void. They did not receive land, means of production, gold or even food for the road. Under those circumstances, many of them "chose" to continue suffering under the yoke of their former owners, who provided some material security in comparison to the adventure of walking the uncertain and hard road to "freedom".

Those who embarked on the adventure of walking to "freedom" saw that once again all they had was their capacity to create living conditions, starting a new cycle of resilience. Contingents of manumitted people arrived in the Pacific, expanding the shelter process that the *cuadrilla*, the Maroons and the *libertos* had already begun. They needed to find places to cook, to sleep, to have a personal and collective life, then conquer and settle them. Countless Black-Africans searching for places to live and to "enjoy" their free status in this way arrived in the region as this fragment from Triana, cited by Escobar, shows us:

> Los negros libres de Barbacoas se establecieron en la costa para gozar de su libertad, lejos de sus antiguos amos, y aquí el mar les fue propicio con sus dones. La multiplicación de la especie ha sido consecuencia del bienestar que encontraon en esta isla y sus contornos (Escobar, 2008: 63).

Without a doubt the wet littoral offered the Black-Africans some kind of wealth. These lands were beautiful and rich in natural resources, and were abundant in fauna and vegetation. Yet the rain forest was hot, wet and hostile with acidic soils; only certain kinds of products could be cultivated there. In addition, the forest was plentiful with wild, fierce vegetation. Thus, the status of liberty mainly meant hard work, with the difference being that this time, the Black-Africans would profit from their work, at least for some time.

Sheltering the Pacific from colonization was for the Black-Afro population a dynamic and changing process. Groups of relatives occupied the riverside and mangroves and organized themselves around kinship relations and the distribution of labor. The settlement was probably a process of trial and error. Some of the issues to be solved concerned the kind of materials and building methods to be employed to build houses, as well as the use and distribution of space and natural and social resources. They must have put into practice the information and knowledge collected by generations over almost three centuries.

The settlement began with the construction of a kitchen with a hearth, which constituted the first movement toward a permanent or temporary settlement. The kitchen was not only a place to cook; in the beginning, it was also the first sleeping room, and the core of the further development of the house (Whitten, 1974). The kitchen was used to store personal products, as well as products for familial and collective tasks such as fishing, farming and hunting. These spaces express, as occurred with the groups of *cuadrillas*, the importance of feeding the group and the process of recuperating energy.

The Pacific littoral zone is rich in fauna and vegetation. The forest, the Pacific Ocean, the mangrove swamps and the large number of lakes and rivers constitute an enormous and varied fountain of food. But to harness the generosity the region offers, the Black-Afro population had to be in constant movement through the littoral in search of those natural resources. Hunting in the forest, fishing in the ocean, lakes and mangroves, and the gathering of wild fruits constitute core activities to maintain the food supply. Those activities were complemented by a production and exchange scheme based on a familial and communitarian distribution of labor. Mixed production units were developed, people became small farmers to cultivate crops such as plantain, rice, beans and fruit for personal and joint consumption, but also collective areas for exploitation, forest, sea, lakes and mangrove swamps, which allowed them to obtain natural resources for their own family consumption, exchange and trade (West, 2005; Whitten & Friedmann, 1974; Leal, 2008).

Together with the activities to meet their material needs, the Black-Afro population needed to fill the social and cultural gap produced by almost four centuries of slavery. By mixing, they built elements and institutions that regulate and guide human behavior. New cultural practices and social institutions emerged as products of the confluence between Hispanic Europe, West Africa and the heat, the humidity, and the wild coast of the Pacific region. The Black-Africans were free; they were no longer subordinated to the control of the white-mestizo institutions and social control regimes. Former slaves and their descendants displayed all their capacities "to sustain and [expand] a population and a culture, in absence of outside support" (Whitten, 1974).

Nevertheless, as was shown in previous pages, the period to enjoy freedom and complete autonomy was short. Just some decades after the manumission laws and the beginning of Black-African settlement in the Pacific lowlands, the agents of the world-system once again arrived in the region.

They were exploring the remote forest in search of natural resources and a labor force to be exploited in order to satiate the demands of the international market and feed the consolidation of the capitalist system. In the late nineteenth century, the lowlands of the Pacific region were once again and definitively inserted into the world-system through continued and repetitive boom-bust cycles. These cycles operated through primary forms of capital accumulation. In every cycle of boom and bust, the fountain of wealth and capital accumulation is constituted on the one hand by the constant environmental production that endowed the Pacific region with conditions to generate natural resources useful as raw materials, and on the other by the existence of a labor force which had not been shaped by wage-labor relations, and had mainly existed outside it without any charge to the capitalist.

The establishment of companies and enterprises brought new social dynamics, particularly in the form of the administration of labor and nature and the use of currency. Contingents of merchandisers coming from remote places arrived in the wet Pacific littoral to expand the global trade networks of raw materials. The new structure of exploitation was no longer based on the old slavery structure and the fervid search for precious minerals, but on the "free market" and the exploitation of other natural resources.

Within the new "free market" scheme, the population had to use much more force for their personal and collective resources and energy in order to collect those products the companies were interested in. The socio-bio-ecological production was materialized in a particular product, then transferred and appropriated by foreigners. In return, nature received nothing, and people received cash to buy goods such as tools, fabrics, and salt at the companies' stores and under the conditions the company set (Leal, 2005, 2008).

Until the middle of the last century, the unequal exchange and the unequal flow of energy and matters occurred periodically, but with intervals of time. The traders focused mainly on resources the region provided without having any particular interest in possessing the land. During the ivory nut trade (1880–1930) and the balsa tree and rubber trade (1935–1944), the labor force and nature were exploited, but without significantly altering the continuity of bio-ecological production and the social metabolism the population developed. In general, Black-Afro groups still controlled the areas where extraction occurred, and the majority of energy was spent on those activities that guaranteed people's autonomous existence. The sale of labor-power was a temporary and sporadic activity done to satisfy the need for resources for social reproduction.

This situation started to change dramatically during the banana cycle (1948–1966), and then with the re-establishment of the timber industry, a renewed gold exploitation era, and the consolidation of new extractive-accumulation activities (shrimp, palm heart, tuna, coconut, cocoa, livestock, and pipelines). Land enclosure began, new technologies were applied in gold mining, non-artisanal fishing and large plantation schemes devastated the forest, poisoned lakes and rivers, and depleted soils. The projects of modernization and development, such as the building of roads, ports and railway, meant the displacement of the inhabitants of the region. An uncontrolled expansion of the urban areas occurred which inherited the social order that had been organized and erected from colonial times and was divided by class and race. "Modern" villages appeared with the comforts of the times, which were inhabited by the new elite – Europeans and white-mestizo families, many of them former slaveholders from the nearby Andes – while afro-descendants lived in simple and precarious hovels (Hoffmann, 1999).

With the help of governments, policy makers, and scientists the interests of the market were imposed. The black population began to be disposed of and blocked away, groups of autonomous farmers and hunters became more "dependent" on a daily wage. The living conditions they had built began to be seriously threatened, and the rift between their existential needs and what they actually received increased. As wage-laborers they earned a misery and there was no longer enough land to work as farmers. Moreover, many lakes and rivers were poisoned while huge industrial fishery companies became "owners" of the ocean, and the plentiful wildlife that had at one time provided food for them was gone. But of course the Black-Afro population was not a passive spectator. Once again, a new and more complex cycle of resilience began.

Third cycle: Upsurge of collective and ethnic identity against exclusion

Modern outdoor parabolic reflector antennas have become part of the landscape of the region Tumaco-San Lorenzo. Almost every house, regardless of how precarious it is, has one. They also have home theater systems, or at least a plasma display. Those appliances are generally acquired by using credit programs which palm oil companies or the outsource entrepreneurs have. Falling into debt today is a constant factor for the population. At times they face the dilemma of either paying debts they have with banks and warehouses or feeding their families.

Since their formation as independent republics, the governments of Colombia and Ecuador, and those sectors with political-economic and cultural power, built and promoted three main ideas, discourses and practices on the Pacific: i) The region is seen as a remote and hostile zone, with a population "distant" and different from the white-mestizo community, as well as from its identity and values; ii) It is promoted as a blessed place with vast natural resources, misused by lazy, but very well adapted *Indios* and "*Negros*"; and iii) It is seen as an area in need of being civilized, colonized, modernized and developed, in order to "free" the population from their moral and material backwardness, and to allow general profiting from the riches the region holds.

This is the framework that has shaped the relationship between people (Black-African and natives) and territories of the Pacific with the states and their respective political and economic administrators and elites. Every boom-bust has operated according to this logic, and the palm oil agribusiness is no exception. Every cycle of exploitation has been promoted either as a tool, means, or opportunity towards the progress of the region. Since the eighties renewed discourses and plans to modernize and develop, both people and nature, have appeared. The region has been transformed into a "development entity" as Escobar (2010) notes, and this project is a national and transnational issue. Private investors work together with governments, development agencies, and international banking institutions such as the IADB and the World Bank.

The new modernizing agenda not only implies changes in the relations both countries have with the region, but also in the way the Black-Afro population here perceive themselves. Thus, along with designing public policies to integrate the region into a transnational trade agenda and constructing the appropriate infrastructure and technologies to intensify the exploitation of nature, there is a configured collective conscience of marginality and exclusion which attempts to neglect and replace Native-Afro settler dynamics (Escobar, 2010). It is true that the population there has been marginalized and become poorer, but in the last decades, the discourse of marginalization and poverty has been intensified and instrumentalized in order to grant legitimacy to neoliberal policies. The nations "need" to cover the debt they have with the Pacific by bringing modernization and development. In "return", people should in many cases abandon their territories, their way of life, their "lack" of rules and unlabored living and subject themselves to the interests of capital and to the will of its technical and administrative staff.

The Black-Afro population contested this exclusive development agenda and its effects by configuring itself as a united and articulated political subject[128] in the nineties by making use of the convulsed political climate and transnational tools, e.g. international conventions. In doing so, black communities emerged as political subjects with their own agenda. Inspired by the indigenous movement, the Black-Afro movement's agenda furthered liberal claims for citizenship and equality. They directly confronted the capitalist expansion which threatens the socio-bio-ecological need for the reproduction of the groups, at the same time denouncing the violence taking place against them: forced displacement, murders, and human rights violations, all of which shape those "development projects".

The communities along the Pacific Coast challenged those ideas, discourses and practices which neglected the long history of the region and which also threaten their socio-bio-ecological reproduction. Farmers, fishermen, and artisanal miners became social activists who, with the help of academics and other political forces, started playing a key role in the new political discourse and praxis in both countries.

Despite the varied composition of this social movement, the main goal consisted in claiming territorial rights and autonomy. Constituted as a collectivity with a common identity, the Black-Afro population confronted the notion of the Pacific as an uninhabited wild place in need of vertical, centralized and external modernizing interventions. Their mobilization around a cultural identity also exposed the weakness of the "racial democracy" and liberal citizenship. Moreover, the political agenda of the Black-Afro

---

128 United and articulated does not mean without conflict.

movement began tearing apart the monopoly and hegemony elites had over the state, forcing the Establishment to recognize alternative conceptions of development not based on the parameters of the market-oriented system and neoliberal policies.

The concrete result of the configuration and mobilization of the Black-African community was Law 70 of 1993 in Colombia, the National Constitution and the Law on Collective Rights of Afro-Ecuadorian Territories in Ecuador. This law, as Baquero notes, "had a progressive character that expanded recognition and clarified property rights for several Afro-descendent communities" (Baquero, 2014; 96). These legal statements grant special rights over the territories where Black-Afro communities have been living and which they have been looking after for more than four centuries:

> Los beneficiarios (las comunidades negras) continuarán conservando, manteniendo o propiciando la regeneración de la vegetación protectora de aguas y garantizando mediante un uso adecuado la persistencia de ecosistemas frágiles, como los manglares y humedales, y protegiendo y conservando las especies de fauna y flora amenzadas o en peligro de extinción (Artículo 21 Ley 70, 1993, Colombia).

The statements also recognized the role the communities have been playing in protecting the environment and biota, the particular production structures the Black-Afro population have, and the involvement of the state in promoting and protecting them:

> La caza, pesca, recolección de productos alimenticios para la subsistencia de los pueblos afroecuatorianos dentro de las tierras en posesión ancestral de las comunidades negras y afroecuatorianas, tendrán prioridad ante el aprovechamiento comercial e industrial (Ley de Pueblos Afroecuatorianos, 2006. Ecuador).

The recognition of the structures of production, circulation, distribution and consumption the Black-Afro communities have maintained for centuries, as well as the state's duty to defend and protect those structures was a milestone in the political sphere in both states orientated by the principles of the market economy. This statute brought the Black-Afro population, at least at the discursive level, social acknowledgment and legal tools to defend the social metabolism they built.

Today, the Black-African communities living in the frontier territories in the Pacific region of Colombia and Ecuador identify themselves, and are recognized by the states, and national and international institutions as an "ethnic group". This recognition, as was illustrated in the second chapter, was the product of a hard struggle carried out mainly by the black rural population in the 1990s in order to attain political rights in defense of their means of living. They have tried to defend the land where they have lived for centuries, the forest, rivers and the lakes that provide them food, and raw materials. They have also tried to defend the ways of life they have built, without any outside support, to guarantee their biological and social reproduction, and to survive a system of death and destruction as was represented by slavery.

The struggle for recognition as an "ethnic group", in the case of Black-Africans in rural communities in Colombia and Ecuador, was mainly concerned with the defense of those systems of knowledge, practices and relations of production, distribution and consumption which guarantee individual and collective existence. People strategically adopted the label "ethnicity", which already had a long political trajectory on the political stage at a national and international level for two reasons. On the one hand, they sought to make visible that the reproduction of the population has been supported by activities that are not guided by the market, the logic of capital accumulation or by the labor-wage relation. On the other hand, they challenged the principles of the market economy by defending the legitimacy of non-alienated production, collective wellbeing and freedom, the latter being understood as maintaining the status of self-determination gained during colonial times.

By appealing to a "collective ethnic identity", groups of autonomous farmers, hunters and fishers on the Pacific coast put "Blackness" on the new political and cultural stage in both countries, gaining at the same time an autonomous administrative structure, *consejos comunitarios* in Colombia and *comunas* in Ecuador. These administrative institutions, despite the problems and contradictions one could see[129], are key as they are a space of formation, cohesion, and interlocution between communities, governments and enterprise as well.

From an historical-materialist perspective, it becomes clear that the notion and experience of Black-African ethnicity are not based on foreign, essentialist, fixed, temporary, or institutionalized notions of "identity" and

---

129 Some problems are the standardization, homogenization and essentialization of Black-Afro identity, which reproduce certain levels of clientelism.

"territory" as some authors such as Escobar (2008), Restrepo (2013) and Hoffmann (2004) have concluded. Instead, the "ethnic identity" articulates and names specific experiences of self-organization and production structures configured by these people for more than four centuries. Along those lines, the "ethnic identity" of Black-Afro communities serves three purposes: first, it connects the experience of those whose personal and collective history was shaped by enslavement; second, it means the whole array of strategies and practices adopted by the Black-Afro population to survive, the particular dynamics of organization they have been developing and maintaining in order to support their material and social reproduction; and third, it represents a political tool in defense of production, food security and social organization, against dispossession, and exploitative projects.

Today, the palm oil agroindustry profits from Black-Afros' capacities of resilience, displayed over five centuries. Mainly descendants of those who survived slavery constitute the labor force on oil palm plantations. People doing "unskilled" labor are the heirs, those who left to fend for themselves, "freed" without any compensation after the manumission laws, who created the material, social and cultural conditions to continue living in the most adverse situation. The richness of palm oil plantations is produced by those who, for more two than centuries, have overcome the socio-ecological destruction and weakness left by every boom-bust cycle of raw material exploitation.

Four centuries of destruction, plundering and spoliation have left behind harmful effects on people and nature. These effects have been exacerbated in the last four decades, when a new colonization race and capitalist expansion began, this time accompanied with technologies never seen before. The search for endless capital profit is generating in Tumaco-San Lorenzo, and also around the world, social and environmental chaos, a metabolic rift which one could say for the case analyzed here that is has had "apocalyptic" implications.

Metabolic rift

Despite the political and administrative divisions that today constitute Tumaco and San Lorenzo, as far as its population is concerned it is a united region. People circulate every day from one city to another, for work, to visit family and relatives, or just for pleasure. The trip could take between two and five hours, depending on whether you travel on foot or by car. In general, people recommend traveling on foot: it is faster and one can also enjoy the beautiful mangrove swamps and avoid the intense sunshine. By car, one can see how the region has become a "green desert". Thousands of homogeneous palms cover an area in which there were huge and legendary trees, and the marvelous wild animals now only exist in peoples' memories. But not only have the flora and wildlife disappeared, so too have the people. In many places there are abandoned structures of what one could assume were once homes.

In the last fifteen years, this region has been a central point of forced displacement. In the late nineties, both sides of the "border" began to suffer the effects of the drug trade: on the one hand, the impacts of "Plan Colombia", which generated re-localization to the Pacific coast of the cocaine trade and the mafia, with their corresponding criminal structure. This, in turn, promoted the establishment of palm oil as the commercial core of Tumaco within the context of the "war" on drugs. On the other hand, the region saw the expansion of the political and military confrontation in Colombia between official and unofficial (paramilitary) armies against the guerrillas. Territories and population were "suddenly" in the middle of a war that Colombia had been suffering for more than six decades. These situations introduced new power interests, who in combination with the classical agents of capitalism changed and disrupted the social metabolism which had been built and maintained by the Black-African population for centuries, and in addition deepened the blockade these groups have suffered.

For more than four centuries, descendants of enslaved Africans have been confronting the different schemes the capitalist system has, to appropriate, control and accumulate surplus value. From the time of colonialism to the present day, every incursion of capitalism in the region has represented a loss through an unequal exchange of matter and energy. Nevertheless, due to the population's capacity for resilience and the natural resources they have, they still remain.

The new cycles of dispossession embodied mainly by the palm oil industry and the context in which it emerged, treated the metabolism created by the bio-ecological structures and the Black-African community as never before. Today, both territories and the population in the Pacific face a process conceptualized by Marx and Engels as a metabolic rift: a situation of socio-ecological destruction associated with the uncontrolled exploitation of natural resources for commercial gain.[130] This metabolic rift is a symptom of the conflict between the methods used to generate material wealth and profit for the market and the production of socio-bio-ecological living conditions.

In region of the Pacific, land has been enclosed and privatized by foreigners, and large mining companies, livestock and trawling have blocked people's access to many products required for their sustenance. Technical developments make environmental and labor exploitation more intensive and faster, reducing the time both have to recover spent matter and energy. Today Tumaco and San Lorenzo are therefore immersed in a social-ecological crisis generated by an historical and constantly renewed process of dispossession and overexploitation of both people and nature. Palm oil synthesizes and condenses this process and its contradictions. The expansion of palm oil plantations risks destroying the bedrock of social resilience, the capacity to keep the group together and to feed it. The agroindustry profits from the capacity for socio-ecological resilience, but at the same time threatens them; this is the biggest contradiction of the capitalist system.

Yet after the crisis caused by the Bud Rot disease, which provoked hunger amongst the population, small farmers renewed their practices of combining crops. In order to guarantee food for all the families, they have confronted the impositions of the technical staff and the requirements of the banks and companies. Together with the struggle for land and the means of production, Black-Afro communities have orientated their struggle towards food sovereignty, which denotes the prevalence of food produced by local communities, as well as local and communitarian control over the supply. Those elements once again occupy center place for the population because, as farmers told me: "With Bud Rot disease we realized that the oil palm can not been eaten".

---

130 Martinez and Bellamy Foster (2000) took up Marx's conceptualization of social metabolism and metabolic rift, attempting not only to rebuild the bridge between Ecology and Marxism; their approach/research gives us a more complex perspective on Marx and Engels' thought, showing us thus that nature and the environment were issues present to the authors' analysis of capitalism.

Conclusions

This chapter was concerned with the questions of the conditions that make the continuity of the process of exploitation and appropriation possible in a "primary form of accumulation" in the region of Tumaco-San Lorenzo. I have argued that it is because of the resilience capacities developed by the environment and the Black-Afro population on the Pacific coast from colonial times to the present day that it has been possible for the capitalist system to find those elements that, along with the palm oil business, constituted the fountain of capital: labor and nature.

I attempted to explain how resilience capacities work. The work has proposed thinking of the Earth as a living system that produces and maintains those conditions that make life possible. This conceptualization provides a framework to show how the elements used within agribusiness – soil, climate, temperatures, sunlight, seed, waters and so on – are a result of complex and constantly renewed interaction between organic and inorganic matter.

The concept social metabolism was introduced to indicate those particular socio-ecological conditions built by the Black-Afro community to guarantee their survival in an unknown and hostile environment such as the Pacific coast. I exposed how enslaved people, Maroons and *libertos* survived and built societies, supported only by their own capacities, "bordered" by societies organized by the colonial structure they inherited, a structure which the Black-African population is still on the lower rungs of. I divided the analysis of the continuing creation process into cycles, taking as a reference point the main moments of boom and bust.

Finally, I argued that the accelerated expansion of extractive business, which implies the intensive use of mechanical and chemical technology, affects the resilience capacities people and the environment have. They generate a metabolic rift, a form of socio-ecological destruction associated with the uncontrolled exploitation of labor and natural resources for commercial gain.

Chapter V. Afterthoughts

Writing an academic work is a hard process. Together with the effort required and the standardized format of the scholarly field, many of us have the goal of and pressure to create some kind of milestone, of developing a new theory or methodology that will change the social sciences. But, after years of hard work and aspiration for innovation, the desire for novelty and change are displaced by haste and the obligation to present a text in an acceptable format.

The hardest part about writing within an academic context has been the feeling of wasting time, resources, energy and money that it generated in me. I must admit that never before have I had such comfortable conditions in which to exercise my passion for reading and writing. Never before have I enjoyed better conditions in which to travel, to meet people around the world, and to share experiences and ideas with them. Despite this great personal experience, I still believe there is an unbalanced exchange occurring between academics and the rest of the world.

I must, for example, think of the millions of trees cut down every year to produce the paper we use to print our texts, dissertations, reviews, magazines, and working papers, many of which are barely read, even within the academic community itself. I think of the millions of CO^2 emissions we produce by traveling to attend conferences, congresses, seminars, and summer or winter schools worldwide. I think about the time many professors and students waste by attending or teaching courses they do not enjoy, as well as the impact the constant competition for resources, jobs and prestige in the academic field have on our physical and mental wellbeing.

The question is, then, why despite those feelings did I decide to enter and continue in academia. There are many reasons. First, for a black, underprivileged woman coming from the lower-class, being part of the academic world was a kind of political struggle. I understood I had a "privileged position" as a social activist that could maybe introduce the closed and select group of people doing an university course to other points of view, other questions, other perspectives. Second, because of the current global context, being a PhD student with a scholarship provided the material conditions to indulge my passion. Third, and possibly the most important reason, researching is one of my passions, and I still believe in the potential scientific knowledge and methodology have to help us to make this world a better place.

This combination of academia and activism encourages me to constantly question every aspect of myself as a researcher: why I wanted to do a PhD, why I chose this specific subject, why a Marxist theoretical framework. I was permanently worried about how to break with what I feel is a generalized idea that Black-Afro academics normally do research on particular issues affecting the Black-Afro population. I asked myself how I could bring my particular experience and concerns as a Black-Afro activist to the academic stage. I wondered what else could be said about the Black-African population, one of the social groups most studied by social scientists, particularly in anthropology, ethnology and cultural studies in Andean countries and Brazil over the last two decades. Was there something new to say about topics such as palm oil and agribusiness, which had already been the subject of thousands of academic texts? How could I explain within a context in which Marxist thought has put aside that it was precisely this framework that provided the best tools to analyze and understand my research problem.

This research was a constant process of self-reflection and confrontation. The more the research progressed, the clearer it became that the pretensions of elaborating new paradigms, new social theories, and new methodologies were just that – pretensions. Many of the ideas I considered new, the product of long periods of reflection, had already been thought and written about. In other words, I realized how narrow the space for novelty is in the academic field. Sometimes, this reality worried me because my work would lack novelty. But, on the other hand, it pleased me to know that the ideas and interpretations I had were reasonable and well-articulated: others had already thought of and understood what I had just discovered.

So, what kind of work could I write in a context where the possibilities of novelty are very limited, in a context where almost everything has apparently been written? Maybe the creativity and originality of social sciences could lie in the capacity researchers have to question certainties and trends, to expose the relationship between the past and the present, to bring different points of view and new questions on topics considered to classic, as well as "classic" perspectives on "new issues", and overall, attempting to return to simple questions. This is what I have tried to do here.

Questioning certainties and trends

I believe that the social sciences have two particular features that represent a major challenge when doing research. On the one hand, we see the permanent emergence of trends and an entire universe of applications of words, concepts, terminology and so on. On the other hand, we have to deal with the rigidity of some words, concepts, and terminology fixed as maxims and certainties, which with time become commonplace, deprived of content and meaning.

Some of the trends my work tried to disassemble or at least to avoid were the "land grabbing" focus, the seductive but empty notion of accumulation by dispossession, and the culturalist perspective predominant in the studies on the Black-Afro population in Latin America. "Land grabbing" is a catch-all term to describe and to study the concentration, legal or illegal, of land and other natural resource mainly by national and transnational companies, but also by governments. The concept "accumulation by dispossession" means the way, according to Harvey, in which capitalism surmounts its current over accumulation crisis. Both these terms have become omnipresent in almost every critical socio-economic research project.

The analysis of Tumaco-San Lorenzo from an historical perspective shows that the palm oil industry is just a new stage in a long struggle between natives, Black-Afros and foreign investors and settlers to control land, natural resources and the labor force. By exploring the history of the palm oil agribusinesses, I have shown that this trade is hundreds of years old, and from the beginning it has exhibited the same characteristics: land enclosure, the dispossession of native communities, the overexploitation of labor, the intensive investment of capital controlled by multinational companies and so on. Thus, what is today conceptualized as land grabbing and accumulation by dispossession is nothing new, but rather has been the basis of capitalist expansion for centuries.

On the other hand, the analysis of the conditions of settlement, and the structures developed by this population to survive in the midst of the most adverse conditions provides me with elements to break with the culturalist focus of the "Black-Afro identity and ethnicity". By analyzing the methods, context and strategies of settlement, "ethnic identity" appears as one strategy used by groups of fishers, farmers, hunters, to defend mainly their means of production, natural resources and autonomous control and administration of labor.

Yet, as I said, trends in the social sciences are one side of "the problem", the other is the very fixed words, concepts and categories. The work of feminist Marxian authors such as Gibson-Graham, Federicci, Picchio and others as well as my own experiences with communities encourage me to rethink concepts like economy, worker, wage-labor, and imperialism. The creative thought of those authors and of people from the Black-African social movement in Central and South America, Europe and the USA, helped me to take apart the idea that capitalism and the wage-labor relations were the core of socio-economic life. Framed by ecological and feminist political economic thought, and the claims of organized Black-African communities, I discovered a universe of communitarian, collective non-profit-oriented production structures that make the existence of human beings possible, structures that paradoxically constitute at the same time the basis for the functioning of capitalism as a world-system.

These complex production structures have been neglected by orthodox, classical and even critical economic analysis. They generate an environment with particular characteristics that make our lives and the existence of humans as biological entities possible; they also produce the resources we use as raw materials and the means of production. Additionally, I attempted to uncover the whole universe of social production responsible for birth, caring, socialization and the teaching of human beings who will become the next labor force.

The main goal of this research was to understand the continuity of the plunder in Tumaco-San Lorenzo. To achieve this, the research had to start by asking simple questions: Where did those natural elements and the labor force needed for the exploitation of palm oil "come from"? What do people and the environment receive in compensation from the palm oil business? By tracking those natural resources needed for a successful plantation, I "discovered", on the one hand, a "universe" of geo-bio-ecological structures of production which function collectively to make our lives as biological beings possible, without any charge to us. I came to understand the planet Earth as a complex, self-regulating system, in which organic and inorganic material interact each other to make possible the existence and the continued reproduction of the environmental features of the planet.

On the other hand, it became clear that the figure of the worker is more an ideological than an analytical tool, used to conceal all of the social relations of production responsible for giving birth, caring for and socializing human beings. Economy has been reduced to the production of means for profit and for capital accumulation while rendering invisible and worthless the complete production relations responsible for social and demographic reproduction. These relations are, in addition, collective and communitarian, not profit-orientated.

Considering the relations between the past and present from a global perspective

I decided to focus my research on the palm oil business in part because of the growing importance of the agribusiness issue has acquired on the academic and political stage. When I started reading the most recent books and articles about this topic, I had the feeling we were, globally, facing a newer and more devastating level of capitalist expansion, which is having particular effects on Tumaco-San Lorenzo. But the more the research went on, the more I understood the current situation with agribusiness in the Pacific region as a constant within the global system. This does not mean that this process is just a simple repetition of the past – it is not. What I learned was that the agribusiness causing such great alarm today has a wider history, and this history must be told. Something similar occurred with the lowlands of the Pacific coast, where land enclosures, dispossession and the overexploitation of labor, as well as the violence this process implies have a longer history.

Thus, the task was to articulate the present and the past of both the palm oil agribusiness and the Tumaco-San Lorenzo region. By studying in-depth the palm oil business, the research exposed the role this product played in the configuration of the world we know today. Palm oil, which replaced the slave trade when it became too expensive or was forbidden, was used to fill and lubricate machinery and railways during the Industrial Revolution, and has been, since the beginning of industrial production, one the most important elements in cleaning chemicals, cosmetics, and food. Furthermore, the palm oil business has been key to the development of powerful industries and multinationals such as Unilever, the United Fruit Company and so on.

In addition, with the help of scientists and policy makers, palm oil has been part of the strategy to develop neocolonialism and imperialism. Now as ever, the development of palm oil plantations and mills implies a huge movement of resources, institutions, knowledge and money at local, national, regional, continental and transcontinental levels. Thanks to the coordinated work of merchants, scientists (mainly botanists and engineers) and policy makers, the palm and the imperialist scheme it belongs to have conquered almost all tropical lands around the planet.

However, as the inhabitants in Tumaco-San Lorenzo say, the palm oil monoculture is just the new face of a larger process. For more than five centuries, the native population and the Black-African settlers have seen thousands of adventurers in search of riches come and go. First there were the Spaniards, looking for El Dorado, then, after the dismantling of the colonial administration and the establishment of the "free market", the English, Italians, and the French arrived in search of other precious metals, raw materials or whatever could be commercialized on the international market. Thus the cycles of nut ivory, rubber, timber, cocoa, banana, and others began and ended. With help from local and national governments, trade exploited nature and labor as intensively as possible, leaving behind neither progress nor development.

Today the palm oil agroindustry, as well as other legal and illegal businesses operating in this region, profit from the historical blockade the population has suffered for more than four centuries. The current socio-ecological crises this region has endured, expressed in high levels of poverty, criminality, and mass displacement, are the product of a long history of exploitation of both humans and nature in order to guarantee capital accumulation; they are produced by an unequal exchange in which the socio-bio-ecological production meant wealth for a few and poverty and destruction for many.

Anyone attempting to understand and explain current conflicts surrounding land, natural-environmental resources and the struggle of native smallholders and local farmers should not lose sight of either the historical or the global perspective. The small village of Tumaco-San Lorenzo, just like almost the whole planet, has been articulated for more than four centuries, to the world-system of accumulation. What occurs today in this small village on the Pacific coast is closely connected with the situation of small farmers in England in the sixteenth century, but also the current situation of small farmers around the world, e.g. in Ethiopia, South Sudan, Ukraine, or Pakistan.[131] There is a close connection between the murders of Black-African leaders in Tumaco and the mass suicides of small farmers in India. There are many common events connecting the increase of homelessness and unemployment in Spain, Greece, Italy and the USA, the so-called refugee crises in Europe and the mass displacement of the Black-Afro population in Tumaco-San Lorenzo. Our task as researchers is to discover them and to make the right connections.

New perspectives of classical issues

Social research moves between the "new" and the "old", between "trends" and "certainties", between "the problems of the past" and "the problems of the present". The issues we decide to study are shaped by the socio-political context we live in, but they are also the product of our personal experience, choices and possibilities. In the same way, discussions and debates in societies and the political agenda could be influenced in some way by scholarly research.

All of these external and internal elements define our research projects. As researchers, we are part of a very closed community and are deeply influenced by what is produced within the academic field. Depending on how open the scholarly community is, how much and how well the interlocution flows with other social groups, our research interests can also be influenced by the agenda of social groups. In this way, social movements have influenced social research on various topics of interest, some of which include governance, security, terrorism, crisis, inequality, race, gender, ethnicity, land conflicts and dispossession.

---

131   Fred Pearce (2012) has written a good summary on current land plundering worldwide.

Some topics constitute a "novelty" in the research camp, others merely reappear, thus exposing the continuance of what one could call "old social problems". The fact that some issues reappear on the academic agenda expresses the permanency of particular "social problems". That is the case of the concept of primitive accumulation, or better put, the continuity of the problem of plunder and dispossession.

For more than one and a half centuries, critical social scientists have continued to rethink and re-conceptualize Marx and Engels' work on why and how capitalist development generates land enclosure, the privatization of commons and large contingents of dispossessed people. The notion of primitive accumulation has, for decades, constituted a key analytical framework within the scholarly debate. From 1867, when <u>Capital</u> was first published, until today, in the middle of an accumulation crisis hitting people worldwide, the continuity of primitive accumulation, the continuity of land enclosure, and the plunder of the means of production in order to generate profit once again constitute a key point in critical social research.

Many past discussions on the nature of the dispossession process have already been overcome, yet hardly a social researcher has argued that primitive accumulation was an initial and finished phase of capitalism. Today, among those studying capitalism from a critical perspective, there is a consensus on the continuance of those processes described and analyzed by Marx and Engels in the fourth chapter of <u>Capital</u>. Researchers from the entire planet are concerned with understanding how, after five centuries of the establishment of the capitalist system, the process that converted farmers and craftsmen into a dispossessed labor force continues today.

Many authors have extended and made the analysis of enclosure and dispossession more complex, exposing that in order for there to be capitalist profit, the global plunder of the environment, culture, cities, identities, etc., is required. Among the many wonderful authors I rediscovered during my research time, I would like to highlight the work of David Harvey, Silvia Federicci and André Gunder Frank. These three authors in particular helped me to understand the whole dimension and the complexity of the plunder.

Despite the gaps in his work, Harvey provided me with some tools to understand key elements in the current geopolitical geography, as well as the close and unbreakable connection between accumulation by expanded production and primitive accumulation. The way in which Harvey takes up Luxemburg's ideas was very clear and useful. He helped me see the dialectical "inside-outside" relation within the capitalist system. He also introduced me to the "universe" of those "pre-existing" external structures, those non-capitalist or non-proletarianized social formations or sectors which capitalist accumulation is dependent on.

Federicci, for her part, gave me an historical and feminist perspective. She explores the early origins of the system, and explains why its consolidation required the material, cultural and political dispossession of women. The result was a crusade against those women who attained political power, material autonomy and knowledge of arts and sciences. Thousands of women were accused of witchcraft, and then they were tortured, hanged or burned. Federici argues that capitalism was born as a counterrevolutionary process. This process implied the alienation of people's means of living, principally of women, in Europe, and then with the colonization of America and Africa. In this way, gender and racial hierarchies were established within the labor force, shaping a dominant European, masculine elite, as well as a modern masculine proletariat.

I was very enthusiastic about both authors and their related perspectives on primitive accumulation. Nevertheless, upon revisiting my research subject, I realized some gaps in the definitions of primitive accumulation that both authors support. I have questions regarding the continuity of primitive accumulation. I did not understand where the pre-existing external factors necessary for current capitalist primitive accumulation and accumulation by dispossession came from. I wondered how, despite more than five centuries of capitalist exploitation in a region such as Tumaco-San Lorenzo, a large percentage of the population is still living from autonomous subsistence farming.

The work of Andre Gunder Frank helped me to find some answers. Frank points out how in the periphery of the system, capitalist accumulation is based on a non-capitalist mode of production, and not wage-labor production relations. For Frank, those other relations of production have three main levels: i) They sustain and guarantee the provision of a potential reserve army of labor and a pool of labor power; ii) They generate the subsistence, sustenance and reproduction of wage-labor for which capital pays less than needed; and iii) They produce the surplus-value that enters into the global capitalist process of capital accumulation.

Frank's optic on the concrete role of non-capitalist productions complimenting wage-labor structures brought me to a new stage in thinking about categories such as labor and others, and to ask simple and basic questions: What is a worker? What is labor? What is production? What is the economy? With the help of political feminist economy and political ecology, it was possible to see, at least partially, the complexity of the universe of relations and structures of production which are not profit-oriented nor regulated by the wage-labor structure, but which make possible and support our existence as human beings. These relations and structures encompass the environmental conditions the planet provides us, as well as the biological features, exchanges of matter and energy that shape us as bodies able to work, and the whole networks of engagement related to our birth, care, socialization and learning.

I also integrated into my analysis the production of environmental, geographical, climatic and ecological conditions so as to have a holistic comprehension of the current global plunder and dispossession. I argued that because of its environmental features, from the colonial period until today, the Pacific region has been a provider of gold and raw materials for the global market. But together with the desired resources, people who are in a condition to extract it are also needed. Without a mass of healthy, skilled and socialized bodies able to work, the "gift" of nature would remain useless.

With these elements as a framework, I realized how much there is still to say about the "old issue" of spoliation. I was sure that despite the many works on agribusiness and the despoiling of the Black-Afro population in the Pacific, my research could bring a new focus to these issues. Thus, my task was to explore, as deeply as possible, the functioning of the palm oil business to show a more complete panorama of capitalism as a system of accumulation in this place, and to explore a more detailed panorama of the spoliation taking place there.

I attempted to move through both the micro and macro level of the palm oil business to expose as completely as possible the socio-bio-ecological structures of production involved in the creation of this commodity. I tried to describe, with the use of basic knowledge, the complex production processes that generate soils, temperatures, rainfall, and water resources needed for a successful palm oil plantation. But the main point was to investigate the origin and maintenance of the labor force. To do so, I focused on the whole production relations that support the existence of the labor force in a context where workers receive from the capitalist less than they need to reproduce themselves and their energies. I argued that the Black-Afro population has been *working to work*. They have not only had to work for the capitalist, but also to reproduce the conditions they need to live. In this way, the labor force has not only had to produce wealth for the capitalists, but also provide for themselves the conditions for reproduction. Thus, the system not only profits from direct exploitation, but and mainly from communitarian, collective and non-proletarian reproductive activities.

… and a "classical perspective" on "new issues"

We are part of the first generation of human beings living and dealing with an imminent global ecological disaster. In the last decades, modern Western industrialized societies have realized the limited capacities the planet Earth has to continue providing those elements that keep the system working and dealing with the waste we produce. These concerns have been materialized in the form of groups, movements and even parties, which appear from time to time hoisting the environmental flag. Over time, most of them have disappeared, others have become the powerful "Green Ecologist Enterprise" claiming the need to take care of the planet, but without challenging the system, and still others are now part of the political state apparatus within which the color "green" is just one more.

I spent more or less six months on the Pacific coast living with small farmers. I observed the effects of chemical and other waste dumped into rivers. I shared with my hosts the fear of drinking the water, of swimming in it, and even of eating the fish caught in the rivers and ocean. I saw large oil marks left behind by industrial fisheries, and saw the oceans and lakes becoming a deposit for plastic garbage. I also heard about the aggressiveness of the fishing trawls razing everything. On that point I understand what an ecological disaster really means, and why a system based on the search for profit is incompatible with the bio-ecological structure this planet holds. I also came to understand many of the tools developed by the Black-Afro population to face continual threats to their survival.

Spending and sharing time with small farmers helped me realize the differences between the "ecological awareness" one has living in cities, where "accountable" is available if you can pay for it, and the "ecological awareness" one has living in the country and forest, where living conditions are dependent on natural-environmental provision. Those topics that today "worry" us around the world constituted the claims of native populations and small farmers long ago. It is they who directly suffer the effects not only of global warming and climate change, but also of the constant spoliation and depletion of nature and people. The fact that global warming particularly affects poor countries and communities (UN, 2007–2008) is the obvious effect of historical spoliation.

I am far from arguing that native populations have a superior environmental conscience. I neither say that small farmers, Afros and the indigenous population have been living in perfect harmony with nature and the environment without destroying it, nor that the capitalist is the only responsible party for all natural devastation. What I am trying to prove, using palm oil as an example is that the functioning of capitalism as a system of accumulation and production is based on the irrational relationship between human beings and nature. It is irrational because its main goal is not to generate human living conditions, but to obtain profit for a small group. Thus, in a system in constant need of expansion, nature and the environment are not only a fountain of raw materials, but are also a commodity for the accumulation of capital. Any use of those elements for other goals is contrary to the internal logic of the system.

Of course, I did not discover anything new. I just acted upon the lack of analysis of those problems we consider "new", e.g. "climate change", "global warming" and "ethnicity" within a framework that studies structures and power relations. For more than one and a half centuries, authors have been exposing the connection between the advance of capitalism and the degradation of the environment. This is not a new issue, but unfortunately it was, for a long time, relegated to the social sciences, and even there only to the critical wing. Even the environmental movement has omitted the connection between environmental crises and a system oriented toward profit.

Today, some authors, activists and politicians are making the right links by placing at the center of the debate on climate change and other "environmental problems", the critique of the system based on the overexploitation of nature and people as Marx, Engels and Luxemburg did. Every phase of capitalist penetration suffered by the Pacific region has been framed by the logic of exploiting nature, the environment and population groups as fast and widely as possible.

The Black-Afro population was introduced to the region as a labor force to exploit the main resource demanded by global trade at the time – gold. Nevertheless, upon their arrival, they had to develop another relationship with the places they now called home. They learned to establish the rational use and exploitation of nature in order to create and guarantee their own conditions of reproduction. Their settlement in the region and exploitation of nature always functioned to provide living conditions for the members of the community, consuming just the amount needed for this task. Thus, by making a rational use of nature, this population built, without any outside support, optimal conditions for the production of communities for more or less four centuries, and kept the geo-ecological features of this place almost "untouched".

Yet, as said before, this rational use of nature is not the product of a mystical or advanced "environmental conscience". It is the product of the survival dynamic that the Black-Afro population had to develop. They have not only had to obtain material elements for their reproduction, but also had to face every phase of spoliation for capital accumulation, which, for them, began with the Transatlantic Trade. Agribusiness, and all the problems and concerns it generates, including the ecological ones, is part of a process that has been performed for almost five centuries.

In this respect, the main goal of this research was to propose a thesis to explain those conditions making the continuity and repetition of the plunder possible. I attempted to develop a framework to explain why, after over five centuries of unequal exchange through colonization and imperialism, capitalists still find nature and laboring people to despoil, in other words, why primary capitalist accumulation is still possible. My thesis is that the Black-Afro population and environment have resilience capacities, which allow them to cope with the constant unequal exchange of matter and energy that characterized the imperialist scheme, in this way overcoming the constant plunder of nature and labor.

Despite my poor and limited knowledge of natural sciences, I attempted to provide a holistic picture of the lowlands of the Pacific coast. I looked briefly at the process of production of those natural elements characteristic of the lowlands on the Pacific coast, and showed how despite centuries of exploitation this place still has wonderful natural-environmental characteristics. Then, I focused on how the Black-African population arrived and settled here, and the conditions under which they became the dominant group and the majority. The text analyzed the relation of the Black-African population with the capitalist system, and organized the process of resilience and spoliation by cycles.

The first cycle was concerned with the slave trade, survival and the search for freedom. Here I pointed out how slaves, Maroons, and the *liberto* population faced the challenge of finding materials and meeting the basic needs for survival as well as building a familial and social structure in the middle of a strange and hostile environment. The wish for freedom was constant, as were the strategies they developed to attain and maintain it. The second cycle dealt with the effects of the manumission laws and the free market. Slave labor was "freed" without any compensation, nor any basic tools for survival. They had no means of production or money, just their capacity to create and provide living conditions for themselves. Under these circumstances, "free" people arrived on the Pacific coast in order to really be free. But the Pacific lowlands are not a "paradise": They are very hot, wet and hostile. The soils here are acidic and thus not very suitable for agriculture, a place plentiful with wild and fiery vegetation. Thus, the status of liberty mainly meant hard work. It was a process of trial and error, not only to organize social life, but also to articulate strategies to face the new spoliation cycles embodied by liberalism and neoliberal schemes.

One of these strategies, dealt with in the third cycle, was social mobilization and raising the flag of an "ethnic identity". The ethnic identity of the Black-Afro community is a tool for the defense of territories, means and modes of production and autonomy. Also, it is a medium through which to propose alternatives to the "horizontal, occidental and capitalist model of development". The emergence of Black-Afro communities and the collective ethnic identity should be understood from an historical and materialist perspective, analyzing the way and the process through which this dispossessed group has been able to build a structure to provide social and material needs. By analyzing the long history of struggle and confrontation, it became clear that "Black-Afro ethnicity" has less to do with the search for "African cultural heritage" than with the defense of a means of production and living conditions.

Within a new cycle of capitalist expansion, specialized in plundering people and nature faster and more "efficiently" than ever before, the ethnic identity of the Black-Afro communities of farmers, fishers and miners must be analyzed as part of an historical struggle which, despite its increasing complexity, has the same bases as long ago, and expresses the confrontation between those who possess a great deal and those who possess little or nothing.

Bibliography

Acevedo, Paola: *Herramientas de análisis de alternativas de producción incorporando el ACV "Cuna a Cuna" a los métodos tradicionales. Comparación de biodisel de palma E. Higuerilla*. Doctoral dissertation. Universidad de Santander, Colombia: Santander. 2012.

Acosta, Misael: *The forests of the northwest of Ecuador*. Report presented to the "Georgia Pacific Co MAS Publicaciones Científicas: Quito. 1969.

Acosta, Misael: *El noroccidente ecuatoriano. Geografía y ecología de Lita a San Lorenzo. Cubierta vegetal y reconocimiento botánico forestal. Agricultura y sugerencias en favor del mejor aprovechamiento de las tierras* (Contribución No. 30). Instituto Ecuatoriano de Ciencias Naturales: Quito. 1959.

Acosta, Misael: *Nuevas contribuciones al conocimiento de la Provincia de Esmeraldas*: Publiciones Científicas: Quito. 1944.

Adams. W, Antonio: *Producción y consumo de aceites vegetales en Colombia*. Centro Interamericano de Reforma Agraria, Instituto Latinoamericano de Ciencias Agricolas OEA: Bogotá. 1965.

Aghalino. S.O.: «British colonial policies and the oil palm industry in the Niger Delta region of Nigeria, 1900–1960». *African Study Monographs*, 21(1), 2000, pp. 19–33.

Agier, Michel «et al»: *Tumaco: Haciendo ciudad*. Universidad del Valle: Cali. 1999.

Aguilera, Maria M.: *Palma africana en la Costa Caribe. Un semillero de empresas solidarias*. Documentos de Trabajo sobre Economía Regional, 30. Banco de la República: Cartagena. 2002.

Akram-Lodhi, A. Haroon: "Land, Markets and Neoliberal Enclosure: An Agrarian Political Economy Perspective." *Third World Quarterly*, 28(8), 2007, pp. 1437–1456., www.jstor.org/stable/20455011.

Alam, M. Anis: "Imperialism and Science." *Social Scientist*, 6(5), 1977, pp. 3–15.

Allen, C. G./Donnithorne, Audrey: *Western enterprise in Indonesia and Malaysia. A Study on Economic Development*. Routledge. Oxfon. 2003.

Almario, Oscar: «Tradición oral e historia oficial en la formación de la identidad de los grupos negros del sur colombiano». Mosquera, Claudia (ed.): *150 años de la abolición de la esclavitud en Colombia*. Universidad Nacional: Bogotá, 2003, pp. 106–151.

Altropico: *Territorios ancestrales, identidad y palma. Una lectura desde las comunidades afroecuatorianas*. Esmeraldas. 2008.

Amin, Samir: *Accumulation on a world scale*. Monthly Review Press: London. 1974.

Andrade, Gerardo: «Aprecio económico y desprecio social del negro». Savoia Rafael (ed.): *El negro en la historia de Ecuador y sur de Colombia*. Centro Cultural Afro Ecuatoriano: Quito. 1988.

Anker, Richard: *Estimating a living wage. A methodological review*. Conditions of work and employment Series. 29. International Labor Organization: Genove. 2011.

Antón, Jhon: *El proceso organizativo afroecuatoriano: 1999–2009*. FLACSO: Quito. 2013.

Arevalo, Diego: *Una mirada a la agricultura de Colombia desde su huella hídrica*. WWF Reporte Colombia: Bogotá. 2012.

Arosemena, Guillermo: *La Gran Bretaña en el desarrollo económico del Ecuador: 1820–1930*. Retrieved 20.04.2013 from http://works.bepress.com/guillermo_arosemena/127/

Arrighi, Giovanni: *The long twentieth century*. Verso. London: 2000.

Arrighi, Giovanni: *El largo siglo XX: Dinero y poder en los orígenes de nuestra época*. Akal: Madrid. 1999.

Arrighi, Giovanni: "Peripheralization of Southern Africa: Changes in Production Processes." *Review Fernand Braudel Center*, 3(2). 1979, pp. 161.

ASAGRAPA: *Agrocombustibles – el negocio agrario*. BASE Investigaciones Sociales. 2008.

Azzelini, Dario: *El negocio de la guerra*. Monte Avila: Caracas. 2009.

Baquero, Jairo: *Layerd Inequalities. Land grabbing, collective rights and Afro-descendant resistance in Colombia*. Policies, Society and Community in a Globalizinh World. 16. LIT Verlag: Münster. 2014.

Barnes, Jon: "Driving roads through land rights: The Colombian Plan Pacífico". *The Ecologist* 23(4), 1993, pp. 135–140.

Batallas, Pablo: *La deforestación en el Norte de Esmeraldas (Eloy Alfaron y San Lorenzo)*. Abya Yala: Quito. 2013.

Bastidas, Silvio / Peña, Eduardo / Reyes, Rafael: «Genealogía del Germoplasma de Palma de Aceite del proyecto de mejoramiento genético de Corpoica». *Revistas Palmas* 24 (1), 2003, pp. 21–29.

Beatty-Medina, Charles: "Maroon chief de Illescas' letter to the Crown, 1586". Joy Kathrym / Garofalo, Leo (eds): *Afro-Latino voices. Narratives from the early modern Ibero-Atlantic world, 1550–1812.* Hackett Publishing Company: Indianapolis, 2015, pp. 20–26.

Berger K, G. / Martin, S, M.: "Oil palm". Kenneth, Kiple / Ornelas, Krimhild (ed.): *The Cambridge world history of the food. Vol. 1.* Cambridge University Press: Cambridge, 2000, pp. 397–310.

Bjornerud, Marcia: "Gaia: Gender and Scientific Representations of the Earth," *NWSA Journal, 9* (3), 1997, pp. 89–96.

Borras, Saturnino / Franco, Jeniffer: "Global land grabbing and trajectories of agrarian change: A preliminary analysis". *Journal of Agrarian Change*, 12(1), 2012, pp. 34–59.

Bouchard, Jean: «Arqueología de la región Tumaco». *Revista Colombiana de Antropología*, XXIV Bogota, 1985, pp. 25–334.

Boyer, Miriam: *Nature materialities and economic valuation conceptual perspectives and their relevance for the study of social inequalities.* Desigualdades Working Paper, 85. 2015.

Braudel, Ferdinan: *La dinámica del capitalismo.* Fondo de Cultura Económica: Ciudad de México. 2006.

Brett, Benet/ Hogde, Joseph. (ed.). *Science and empire: Knowledge and networks of science across the British Empire, 1800–1970.* Britain and the world Series. Palgrave Macmillan: Hampshire. 2013.

Buechler, Stephanie / Hanson, Anne-Marie (ed.): *A political ecology of women, water and global environmental change.* Routledge International Studies of Women and Place Series. University of California, Davis and University of Arizona: Arizona. 2015.

Buitron, Ricardo: «El caso de Ecuador y el paraíso en siete años. El amargo fruto de la palma aceitera: Despojo y deforestación». Retrieved 18.05.2011 from http://wrm.org.uy/oldsite/plantaciones/material/palma3.html

Bunker, Stephen G.: "Modes of Extraction, Unequal Exchange, and the Progressive Underdevelopment of an Extreme Periphery: The Brazilian

Amazon, 1600–1980." *American Journal of Sociology*, 89(5), 1984, pp. 1017–1064.

Burgos, Roberto: *Rutas de libertad: 500 años de travesía*. Ministerio de Cultura, Universidad Pontificia Javeriana: Bogotá. 2010.

Burgos, Carlos: *Medios de vida adaptados por los habitantes del núcleo productivo de la palma de aceite en la vereda Imbil, Municipio de Tumaco, zona occidental de Colombia, como respuesta a la crisis económico y social causada por la enfermedad. La Pudrición del Cogollo (PC)*. Master thesis, Universidad Pontificia Javeriana: Bogotá. 2013.

Cañas, Verónica: *Conflicto socio ambiental y laboral entre la comunidad Carondelet y la palmicultora Palmeras del Pacífico Actores plurales y diversas miradas*. Master's thesis, Facultad Latinoamericana de Ciencias Sociales, Ecuador: Quito. 2009.

Carrión, Lucia / Cuvi, Maria: *La Palma africana en el Ecuador: Tecnología y expansión empresarial*. Facultad Latinoamericana de Ciencias Sociales (FLACSO) Colección de Investigaciones, 4: Quito. 1985.

Castañeda Andrés / Flores Javier: *Aplicación del análisis de ciclo de vida para la producción de biodisel a partir del aceite de palma empleando la metodología "de la cuna a la cuna"*. Master tesis, Universidad Industrial de Santander, Colombia: Santander. 2008.

Castro, Alfonso. «Revolución de Concha». Savoia Rafael (ed.): *El negro en la historia del Ecuador y sur de Colombia*. Centro Cultural Afro-Ecuatoriano: Quito. 1988, pp. 31–60.

Céspedes, Guillermo: *América Hispánica*. Editorial Labor: Barcelona. 1983.

Charvet, Erica: *Feminidad y Masculinidad en la cultura Afroecuatoriana*. Abya-Yala: Quito. 2010.

Chesworth, Ward (ed.): *Encyclopedia of soil sciences*. Springer: Dordrecht. 2008.

Christopher, A. J.: "Patterns of British overseas investment in land, 1885–1913". *Transactions of the Institute of British Geographers, New Series*, 10(4), 1985, pp. 452–466.

Clare, Patricia: "Poder y medio ambiente. La palma aceitera en el Pacífico Costarricense, 1950–2007". *Revista de Agricultura e Historia Rural*, 57. 2012, pp. 135–166.

Colchester, Marcus / Chao, Sophie (ed.): *Oil Palm expansion in South East Asia. Trends and implications for local communities and indigenous people*. Forest People Program. 2013.

Colmenares, German: *Historia económica de Colombia*. Universidad del Valle: Cali. 1979.

Corley, RHV / Tinker, B: *The oil palm*. Blackwell Science Ltd: New Jersey. 2003.

Corner, Edred: *The natural history of the palms*. Universty of California Press: California. 1966.

Corsetti, Giancarlo / Motta, Nancy / Tassara, Carlo: *Cambios Tecnológicos, organización social y actividades productivas en la costa pacífica colombiana*. Comitato Internazionale Per Lo Sviluppo Dei Popoli: Bogotá. 1990.

Crosby, Alfred: *Ecological imperialism: The biological expansion of Europe, 900–1900*. University of Texas. Cambridge Press: Austin. 2004.

Cross, Michael: "Collapse of the great oil palm bubble". *New Scientists*, October. 108(1480), 1985, pp. 17.

De Angelis, Massimo: "Marx and primitive accumulation: The continuous character of capital's "enclosures". *The Commoner*, 2 (Enclosures: The mirror imagen of alternatives) 2001. Retrieved 20.08.2013, from http://www.commoner.org.uk/02deangelis.pdf.

Dodson, Callaway / Gentry Alwyn: "Extinción biológica en el Ecuador occidental". Mena, Patricio / Suarez Luis (eds.): *La investigación para la conservación de la diversidad biológica en el Ecuador*. Quito: EcoCiencia. 1993, pp. 27–60.

Dörre, Klaus: "Social classes in the process of capitalist landnahme on the relevance of secondary exploitation". *Socialist Studies / Études socialistes*, 6(2), 2010, pp. 43–74.

Dunaway, Wilma: "The double register of history: Situating the forgotten woman and her household in capitalist commodity chains". *Journal of World-System Research*, 7, 2011, pp. 2–31.

Elmhirst, Rebecca. "Introducing new feminist political ecologies". *Geoforum*, 42(2), 2011, pp. 129–132.

Engels, Friedrich: *Dialektik der Natur*, S. 2. Digitale Bibliothek Band 11: Marx/Engels, S. 8321f (vgl. MEW Bd. 20, S. 307f).

Escobar, Arturo: *Territorios de diferencia. Lugar, movimiento, vidas y redes*. Departamento de Antropología. Universidad de Carolina del Norte. Samava Impresiones: Popayan. 2010.

Escobar, Arturo (ed.): *Pacífico: ¿Desarrollo o diversidad? Estado, capital y movimientos sociales en el Pacífico colombiano*. CEREC, Ecofondo: Bogotá. 1996.

Estupiñan, Julio: *Volviendo a vivir*. Letra Mía, ED: Quito. 2003.

Estupiñan, Julio: *Banano amargo*. Fondo de la Casa de la Cultura Ecuatoriana. Quito. 1994.

Estupiñan, Julio: *Los Estupiñan Tello en el Ecuador*. Casa de la Cultura Ecuatoriana: Quito. 1991.

Federici, Silvia: *Revolución en punto cero: Trabajo domestico, reproducción y luchas feministas*. Traficantes de Sueños: Madrid. 2013.

Federici, Silvia: *Revolution at point zero: House*work, reproduction, and feminist struggle. Pm Press: Oakland. 2012.

Federici, Silvia: *Caliban y la bruja. Mujeres, cuerpos y acumulación originaria*. Traficantes de Sueños: Madrid. 2010.

Federici, Silvia: "Debt crisis, Africa and the new enclosures". Midnight Notes, The New Enclosure. (10), 1990, pp. 10–17.

Fenstermaker, Sarah (ed.): *Women and household labor*. Plenum. New York: 1979.

Fernandez, Paloma: *Afrodescendencia en el Ecuador. Raza y género desde los tiempos de la Colonia*. Quito: Abya-Yala: Quito. 2001.

Fernandez, Paloma: *Diáspora Africana en América Latina: Discontinuidad racial y maternidad política en Ecuador*. Doctoral dissertation, Universidad del País Vasco: Bilbao. 1999.

Fieldhouse, David: *Unilever overseas. The anatomy of an multinational 1895–1965*. Hoover Institution Press publication: London. 1978.

Filomeno, Felipe: *Monsanto and the intellectual property in South America:* Palgrave Macmillan: New York. 2014.

Fischer, Geoffrey: "Conflict in the Pailon: The British experience in Esmeraldas Provence, Ecuador, 1860–1914". Marshall, Oliver (ed.): *English speaking communities in Latin America*. Macmillan: London, 2002, pp. 105.

Florez, Telmo: *Tumaco reseña histórica*. Editorial Pacífico: Cali. 1983.

Foster, Jhon Bellamy: *Marx's Ecology: Materialism and Nature*. Monthly Review Press. New York. 2000.

Foster, Jhon Bellamy: Marx's theory of metabolic rift: Classical foundations for environmental sociology. *American Journal of Sociology*, 105(2), 1999, pp. 366–405.

Foster, Jhon Bellamy / Clark, Bret: "Ecological imperialism and the global metabolic rift: Unequal exchange and guano-nitrate trade". *International Journal of Comparative Sociology* June/August 2009, 50 (3–4), 2009, pp. 311–334.

Foucault, Michael: *Language, counter-memory, practices*. Cornell University Press: New York. 1977.

Foucault, Michael: *The Birth of Biopolitics*. Palgrave Macmillan: Hampshire. 2008.

Folbre, Nancy: *Valuing children rethinking the economics of the family*. Harvard University Press: Cambrigde. 2008.

Folbre, Nancy: *Who pays for the kids. Gender and the structures of constrain*. Routlege: New York. 2003.

Forero, Jaime: *Economía campesina y sistema alimentario en Colombia: Aportes para la discusión sobre seguridad alimentaria*. Facultad de Estudios Ambientales y Rurales, Universidad Javeriana: Bogotá. 2003.

Frank, Andre Gunder: *Capitalism and underdevelopment in Latin America: Historical studies of Chile and Brazil*. Monthly Review Press: London and New York. 1982.

Frank, Andre Gunder: *The world accumulation: 1492–1789*. Monthly Review Press: London. 1978.

Frank, Andre Gunder: "Industrial capital stocks and energy consumption". *The Economic Journal*, 69(273), 1959, pp. 170–174.

Friedmann, Nina / Arocha, Jaime: *De sol a sol, genésis, transformación y presencia de los Negros en Colombia*. Fangin Books. Chicago. 1986.

Garcia, Jaime: *Territorios, territorialidad y desterritorialización. Un ejercicio pedagógico para reflexionar sobre los territorios ancestrales*. Fundación Altrópico: Quito. 2010.

Gentry, Alwyn: «Northwest South America (Colombia, Ecuador and Perú)». Campbell, David/ Hammond, David (eds.): *Floristic inventory of tropical countries: The status of plant systematics, collections, and vegetation, plus recommendations for the future.* Bronx, USA: New York Botanical Gardens. 1989, pp. 137–160.

Gentry, Alwyn: "Species richness and floristic composition of Chocó region plant communities". *Caldasia*, 15, 1986, pp. 71–91.

Gibson-Graham, J. K.: "Beyond global vs local: Economics politics outside the binary frame in forthcoming". Herod, Andrew / Wright Milissa (eds.): *Geographies of power: Placing scale.* Blackwell: Oxford. 2002, pp. 35–60.

Gibson-Graham, J. K.: *Una política postcapitalista.* Siglo del Hombre Editores, Pontificia Universidad Javeriana, Facultad de Piscologia, Instituto Pensar: Bogotá. 2001.

Gibson-Graham, J. K: *The end of capitalism (as we knew it).* Blackwell: Oxford. 1996.

Grandia, Liza. *The tragedy of enclosures: Rethinking primitive accumulation from the Guatemalan hinterland.* Paper presented to the spring colloquium at the Program in Agrarian Studies Yale University. 2007.

Gondard, Pierre / León, Juan B. / Slyva, Paola: *Transformaciones agrarias en el Ecuador.* Centro Ecuatoriano de Investigación Geográfica: Quito. 1983.

Guerra de la Espriella, Antonio: «Decada 1980. La década dorada de la palma». *Revista Palmas.* 3, 2002, pp. 93–96.

Guerrero, Andrés: *Los oligarcas del Cacao. Ensayo sobre la acumulación originaria en el Ecuador: hacendados cacaoteros, banqueros, exportadores y comerciantes en Guayaquil 1890–1910.* Editorial el Conejo: Guayaquil. 1983.

Gupta, Surinder (ed.): *Technological innovations in major world crops, Vol. 1.* Springer: New York. 2012.

Habermas, Jürgen: "Technology and science as ideology". Habermas, Jürgen (ed.): *Toward a rational society. Students, protest, science and politics.* Cambridge Poly Press. 1987, pp. 137–160.

Hardt, Michael / Negri, Antonio: *Empire.* Harvard University Press: Massachusetts. 2000.

Hartley, C.W.S.: *The oil palm.* Longman: London. 1988.

Harvey, David: "The crisis and the consolidation of class power. Is this really the end of neoliberalism?". *Counter punch*. 2009. URL: http://davidharvey.org/2009/03/the-crisis-and-the-consolidation-of-class-power

Harvey, David. *The limits to capital*. Verso: London. 2006.

Harvey, David: *Spaces of Global Capitalism: A Theory of Uneven Geographical Development*, Verso: London. 2006.

Harvey, David: *Neoliberalism*. Oxford University: London. 2005.

Harvey, David: *The new imperialism*. Oxford University: New York. 2003.

Headrick, Daniel, R.: *The tentacles of the progress. Technology transfer in the age of imperialism, 1850–1940*. Oxford University Press: Oxford. 1999.

Henderson, J. / Osborne, D.: " The oil palm in all our lives: How this came about". *Endeavour*, 24(2), 2000, pp. 63–80.

Hernandez, Luis / Castro Hernando / Barba, Fabio: *Petróleo y conflicto en el gobierno de la seguridad democrática, 2002–2010*. Instituto de Investigaciones UNIJUS, Facultad de Derecho, Ciencias Políticas y Sociales, Universidad Nacional de Colombia: Bogotá. 2013.

Hochshild, Adam: *King Leopold's ghost: A story of greed, terror and heroism in colonial Africa*. Macmillan: London. 1998.

Hoekstra, A. Y.: "Human appropriation of natural capital: A comparison of ecological footprint and water footprint analysis". *Ecological Economics*, 68 (7), 2008, pp. 1963–1974.

Hoffman, Odile: *Comunidades Negras en el Pacífico Colombiano. Dinámicas e innovaciones étnicas*. Instituto Frances de Estudios Andinos: Lima. 2004.

Hoffmann, Odile: "Sociedades y espacios en el litoral Pacífico sur colombiano". Agier, Michel / Álvarez, Manuela / Hoffmann, Odile / Restrepo, Eduardo: *Tumaco: haciendo ciudad. Historia, cultura e identidad*. Bogota: Ican-Ird-Universidad del Valle: Cali. 1999, pp. 15–53.

Holdridge, Leslie: *Life zone ecology*. Tropical Science Center: San José. 1967.

Houben, Vincent / Seibert, Julia: "(Un) freedom: Colonial labor relations in Belgian Congo and the Netherlands Indias compared". Frankema Ewout / Buelens Frans (eds.), *Colonial exploitation and economic development*. Routledge: Oxford. 2003, pp. 178–92.

Houghton, Juan (ed.): *La tierra contra la muerte. Conflictos territoriales en los pueblos indígenas de Colombia*. Centro de Cooperación al Indígena (CECOIN). Ediciones Antrophos: Bogotá. 2008.

Hovorka, Alice: "The No. 1 ladies' poultry farm: A feminist political ecology of urban agriculture in Botswana". *Gender, Place and Culture*, 13(3), 2006, pp. 207–255.

Kalmanovitz, Salomón: «Consecuencias económicas del proceso de independencia en Colombia». *Revista Economía Institucional*. 10(19), 2008, pp. 217–233.

Kirchberger, Ulrike: „Die "Ecuador Land Company": Ein deutsch-britisches Kolonisationsprojekt in der Mitte des 19. Jahrhunderts". Kirchberger Ulrike: *Aspekte deutsch-britischer Expansion: Die Überseeinteressen der deutschen Migranten in Großbritannien in der Mitte des 19. Jahrhunderts*. Franz Steiner Verlag: Stuttgart. 2013, pp. 166–196.

Kircher, Jhon: *Tropical ecology*. Princeton University Press: New Jersey. 2013.

Klein, Herbert / Ben, Vinson III: *Historia mínima de la esclavitud*. El Colegio de México, Centro de Estudios Históricos: Ciudad de México. 2013.

Kollontai, Alexandra: *The labour of women in the evolution of the economy*. Retrieved 20.09.2014, from https://www.marxists.org/archive/kollonta/

Kupper, A. / Pacheco, R. / Pena, H: *Mision para el estudio y reconocmiento de los suelos de la región Noroccidental del Ecuador*: Quito. 1957.

Janick, Jules / Paull, Robert: *Encyclopedia of fruits and nuts*. Cambridge University Press: Cambride. 2006.

Jurado, Fernando: *Esclavitud en la Costa Pacífica, Isquandé, Tumaco y Barbacoas*. Centro Abya-Yala: Quito. 1990.

Leal, Claudia: «Disputas por taguas y minas: recursos naturales y propiedad territorial en el Pacífico colombiano; 1870–1930». *Revista Colombiana de Antropología*, 42(2), 2008, pp. 409–438.

Leal, Claudia: «Un puerto en la selva. Naturaleza y raza en la creación de la ciudad de Tumaco, 1860–1940». *Historia Crítica,* 30, 2005, pp. 39–65.

Leal, Claudia / Restrepo, Eduardo: *Unos bosques sembrados de aserríos. Historia de la extracción maderera en el pacífico colombiano*. Instituto Colombiano de Antropología es Historia (ICANH), Universidad Nacional de Colombia: Bogotá. 2003.

Lenin, Wladimir: *Imperialismo. La Fase superior del capitalismo*. Fundación Federico Engels. Taurus: Madrid. 2008.

Lenin, Wladimir: *El Desarrollo del capitalismo en Rusia*. Editorial Ariel: Barcelona. 1975.

Ó Loingsigh, Gearóid: *La reconquista del Pacífico: invasión, inversión, impunidad*. Proceso de Comunidades Negras, Coordinador Nacional Agrario: Bogotá. 2013.

Lopez, Eduardo / Callot, Jean-Yves / Sosson, Marc: *Sedimentary constraints on the tectonic evolution of the paired Tumaco–Borbón and Manglares forearc basins (southern Colombia – northern Ecuador) during the late Cenozoic*. Paper presented at the 7th International Symposium on Andean Geodynamics in Nice. 2008.

Lovelock, James: *The vanishing face of Gaia*. Basic Books: New York. 2009.

Lugard, Frederick, D: "The White man's task in Tropical Africa". *Foreign Affairs*, 5(1), 1926, pp. 57–65.

Lukács, Georg: *A defense of "History and class consciousness": Tailism and the dialectic*. Verso: London. 2000.

Luna, Alfredo: *El Estado Ecuatoriano y la violación de los derechos colectivos y ambientales en San Lorenzo, Esmeraldas*. Pontificia Universidad Católica del Ecuador. 2002.

Luxemburg, Rosa. *The accumulation of capital*. Routledge Classics: New York. 2003.

Lynn, Martin: *Commerce and economic change in West Africa: The oil palm trade in the nineteenth century*. Press Syndicate of the University of Cambridge: Cambridge. 1997.

Marchal, Jules: *Lord Leverhulmes's ghost: Colonial exploitation in the Congo*. Verso: New York. 2008.

Mariátegui, Jose: *Ideología y política*. Editorial Popular: Lima. 1985.

Mariátegui, Jose: *Siete ensayos sobre la realidad peruana*. Editorial Popular: Lima. 1984.

Marin, Ruy Mauro / Milla, Margara (ed.): *La teoría social latinoamericana. La centralidad del Marxismo. Vol. III*. Paper presented at Seminario Interno Permanente del Centro de Estudios Latinoamericanos de la Facultad de Ciencias Políticas y Sociales de la Universidad Nacional Autónoma de México. 1995.

Martin, Susan: *European plantations firms and Malaysia's new economic policy since 1970*. Paper presented at IEHA Congress, Session 94: Foreign Companies and Economic Nationalism in the Developing World after II World War. 2006.

Martin, Susan: *The UP Saga*. Nordic Institute of Asian Studies: Denmark. 2003.

Martinez-Alier, Joan: "Marxism, social metabolism and ecologically unequal exchange" Departament d'Economia i d'Història Econòmica, Universitat Autònoma de Barcelona. Retrieved 30.05.2014, from http://www.h-economica.uab.es/papers/wps/2004/2004_01.pdf

Martinez-Novillo, Javier: "Geneología y discurso; de Nietzsche a Foucault". *Nómada. Revista Crítica de Ciencias Sociales y Jurídicas*. 26(2), 2010. pp. 45–60.

Marx, Karl: *Capital. A Critique of Political Economy*. Volume I. Trans. Ben Fowkes. Penguin Books: London. 2004.

Mazzadra, Sandro: "How many histories of labor." *Postcolonial Studies*, 14(2), 2013, pp. 151–170.

Mészáros, István: *Beyond Capital: Towards a theory of transition*. Merlin Press: London. 1995.

Micha, Renata /Khatibzadeh, Shahab / Shi, Peilin / Fahimi, Saman «et al»: "Global, regional, and national consumption levels of dietary fats and oils in 1990 and 2010: a systematic analysis including 266 country-specific nutrition surveys BMJ". 2014, vol 348, pg g2272.

Mideros, Mayra: *Las empresas palmicultoras y la generación de desarrollo económico local en el Cantón San Lorenzo, Provincia de Esmeraldas, 1998–2008*. Master's thesis, Desarrollo y Territorio, Facultad Latinoamerica de Ciencias Sociales, Sede Ecuador: San Lorenzo. 2010.

Mignolo, Walter: "Four pillars of moveable memories: Abya Yala, Europe and Africa in the imperial invention of América and Latino-a América in the US. Conference at the Society Latin America Studies, Manchester. 2013.

MINGA & FUNDESVIC: *Las familias trabajadores de la palma contamos nuestra historia. Memorias de las víctimas del sur del César. Cartilla 1*. FUNDESVIC, Fundación MINGA y Sintraproaceites: Bucaramanga. 2011.

Mingorance, Fidel: *El Flujo del aceite de palma Colombia-Bélguica/Europe. Acercamiento desde una perspectiva de derechos humanos.* Cordination Belge pour Colombie: Bruxelles. 2006.

Moll, H.A.J.: *The economics of oil palm. Economics of crops in developing countries.* No. 2. Centre for Agricultural Publishing and Documentation Wageningen: Wageningen. 1987.

W. Neil, Argen: "Social and ecological resilience: are they related?". *Progress in Human Geography*, 24(3), 2016, pp. 347–364.

Nelson, Julie: *Feminism, objectivity and economics.* London. Routledge: New York. 1996.

Nelson, Julie: *Economics for humans.* University of Chicago Press: Chicago. 2006.

Ngando-Ebongue F. G. "et al": "The oil palm". Gupta, S.K. (ed.): *Technological Innovations in Major Oil Crops.* New York. Springer. 2012, pp. 165–200.

Nuñez, Ana Maria: «Los efectos en la salud humana y el ambiente por la producción de aceite crudo de palma africana en San Lorenzo. Esmeraldas». Altropico: *Territorios ancestrales, identidad y palma. Una lectura desde las comunidades afroecuatorianas.* 2007, pp. 123–150.

Obahiagbon, Fr, I.: "Aspects of the African Oil Palm (Elaeis guineesis jacq.) and the Implications of its Bioactives in Human Health". *American Journal of Biochemistry and Molecular Biology,* 2. 2012, pp. 106–119.

Onffroy de Thoron, Enrique: *América Ecuatorial. Su historia pinteresca y política.* Colección Ecuador. Testimonio de Autores Extranjeros. Editora Nacional: Quito. 1983.

Ortiz, Ruben / Fernandez, Olman: *Cultivo de la palma aceitera.* EUNED: San José. 2000.

Pearce, Fred: *Land Grabbing: Der Globale Kampf um Grund und Boden.* Kunstmann Verlag: München. 2012.

Phillips, O. / Lewis, S. L. / Baker, T. R. / Malhí, Y.: "The response of South American tropical forest to contemporary atmospheric change". Bush, Mark / Flenly, Jhon (eds.): *Tropical rainforest responses to climatic change.* Springer: New York. 2007, pp. 317–332.

Picchio, Antonella: «Condiciones de vida: Perspectivas, análisis económico y políticas públicas». *Revista de Economía Crítica*, 7. 2007, pp. 27–54.

Picchio, Antonella: «Visibilidad analítica y política del trabajo de reproducción social». Carrasco, Cristina (ed.): *Mujeres y economía. Nuevas perspectivas para nuevos y viejos problemas.* Editorial ICARA: Barcelona. 2003, pp. 201–243.

Picchio, Antonella: *Un enfoque macroeconómico "ampliado" de las condiciones de vida.* Paper presented at the Conferencia Inaugural de las Jornadas "Tiempos, trabajos y género", Facultad de Ciencias Económicas de la Universidad de Barcelona. 2001.

Pleanjai, Somporn / Geehlawa, Shabbir/ Garivait, Savitri: "Environmental evaluation of biodiesel production from oil palm in a life cycle perspective". *Asian Journal on Energy & Environment*, 8(1,2), 2007, pp. 15–32.

Polanyi, Karl: *The great transformation: The political and economic origins our time.* Beacon Press: Boston. 1958.

Potter, Lesley: «La industria de aceite de palma en Ecuador: ¿un buen negocio para los pequeños agricultores». *Eutopia, Revista de Desarrollo Económico y Territorial*, 2. 2011, pp. 39–54.

Poulantzas, Nico: *Las clases sociales en el capitalismo actual.* Siglo XXI Editores: Ciudad de México. 2007.

Poulantzas, Nico: *Poder político y clases sociales en el estado capitalista.* Siglo XXI Editores: Ciudad de México. 1972.

Pye, Oliver / Bhattacharya, Jayati (eds.): *The oil palm controversy in Southeast Asia, transnational perspective.* ISEAS Publishing: Singapur. 2013.

Quevedo, Norbey / Laverde, Juan David: El Dossier de los Palmeros. *El Espectador*, Bogotá, 20 Jan. 2008. Retrieved 19.07.2013 from http://www.elespectador.com/impreso/investigacion/articuloimpreso-3695-el-dossier-de-los-palmeros

Quijano, Anibal: «Colonialidad del poder, eurocentrismo y América Latina». Lander, Edgardo (ed.): *La colonialidad del saber: eurocentrismo y ciencias sociales: Perspectivas Latinoamericanas*, CLACSO, Consejo Latinoamericano de Ciencias Sociales: Buenos Aires. 2000, pp. 201–246.

Quintero, Leonardo: *Estudios biométricos aplicados a la selección de familias Tenera para la obtención de pisiferas progenitora.* Universidad de Guayaquil, Ecuador: Guayaquil. 2010.

Randeria, Shalini: *Geteilte Geschichte und verwobene Moderne.* "Das" Arab. Buch: Berlin. 1999.

Rangel, Alfredo / Ramirez, Willliam / Betancurt, Paola: *La palma africana: Mitos y realidades del conflicto*. Fundción Seguridad y Democracia: Bogotá. 2009.

Rangel, Alfredo: «Conflicto social y zonas palmeras en Colombia». Revista Palmas, 29(2), 2008, pp. 45–49.

Fernádez, Paloma: *Diáspora africana en América Latina: Discontinuidad racial y meternidad política en Ecuador*. Doctoral dissertation, Universidad del País Vasco: Bilbao. 2009.

Reid, Jhon: *Warrior aristocrats in crisis: The political effects of the transition from the slave trade to oil palm commerce in the nineteenth century kingdom of Dahomey*. Doctoral dissertation, University of Stirling, Scotland. 1986.

Restrepo, Eduardo: *Etnización de la negridad: La invención de las 'comunidades negras' como grupo étnico en Colombia*. Observatorio de Territorios Étnicos Editorial Universidad del Cauca: Popayan. 2013.

Restrepo, Eduardo: *De 'refugio de paz' a la pesadilla de la guerra: Implicaciones del conflicto armado en el proceso organizativo de 'comunidades negras' del Pacífico nariñense*. Informe División de Antropología Social Instituto Colombiano de Antropología e Historia: Bogotá. 2005.

Restrepo, Eduardo / Cortez, Hernán: *Deforestación y degradación de los bosques en el territorio-región de las comunidades negras del Pacífico colombiano*. Retrieved, 20.08.2012, from: http://wrm.org.uy/oldsite/deforestation/LAmerica/Colombia.html

Richardson, D. L.: "The history of oil palm breeding in the United Fruit Company". *ASD Oil Palm Papers*, 11. 1995, pp. 1–22.

Ridler, N, B.: "A case of study in import substitutions: oil palm in Colombia". *Interamerican Economics Affairs*, 29(4), 1976, pp. 21–34.

Rincón, Luis / Cuesta, Andrea / Felix, Erika: *Cálculo del cambio en emisiones generadas asociadas a la expansión de cultivos de palma aceitera en Colombia*. FAO Colombia. Retrieved 17.05.2103, from http://www.fao.org/fileadmin/templates/ex_act/pdf/ex-act_applications/Report-biofuel-colombia.pdf

Rivera, V. Frey / Centro Andino de Acción Popular y Organización Campesina Muisne-Esmeraldas: *Campesinado y organización en Esmeraldas*. Centro Andino de Acción Popular, Organización Campesina Muisne-Esmeraldas: Quito. 1986.

Roa, Ivan: *El desborde de la violencia: Raza, capital y grupos armados en la expansión transnacional de la palma aceitera en Nariño y Esmeraldas.* Master thesis, Facultad Latinoamerica de Ciencias Sociales, Sede Ecuador: Quito. 2012.

Robin, Marie-Monique: *The world according to Monsanto: Pollution, corruption and the control of our food supply.* The New Press: New York. 2008.

Rodriguéz, César «et al»: *Primer informe sobre raza y racismo en Colombia. (Serie Justicia Global).* Bogotá: Universidad de los Andes: Bogotá. 2008.

Romero, Mario: «Familia afrocolombiana y construcción territorial en el Pacífico sur, siglo XVIII». Arocha, Jaime(ed.): *Geografía humana de Colombia. Los afrocolombianos. Tomo 4.* 2004. Retreived 30.03.103, form: http://www.banrepcultural.org/blaavirtual/geografia/afro/familia

Romero, Mario: *Poblamiento y sociedad en el Pacífico colombiano siglos XVI a XVIII.* Universidad del Valle Editorial: Cali. 1995.

Rosengarten Jr., Frederick: *Wilson Popenoe: Agricultural explorer, educator, and friend of Latin America.* National Tropical Botanical Garden: Hawai. 1992.

Rueda, Rocio: *Zambaje y autonomía. Historia de la gente negra en la provincia de Esmeraldas.* Colección Marajeda No. 1: Provincia de Esmeraldas. 2001.

Ruggiero, Vicenzo: *The crimes of the economy. A criminalogical analysis of economic thought.* Routlegde. 2013.

Schiebinger., Londa: *Plants and Empire: Colonial bioprospecting in the Atlantic World.* Harvard University Press: Cambridge. 2004.

Shiva, Vandana: *Ecology and the politics of survival. Conflicts over nature resources in India.* The United States University Press: Tokyo. 1991.

Shiva, Vandana: *Staying alive: Women, ecology and development.* Zed Books: London. 1989.

Soto, Martha «et al» (eds.): *El Poder para ¿qué?: las plantaciones de palma africana, los regímenes del terror, el cartel de la gasolina, empresas e inversiones, cultivos de coca.* Editorial Intermedio: Bogotá. 2007.

Suárez, Mario: «Geological Framework of The Pacific Coast Sedimentary Basins, Western Colombia». *Geología Colombiana,* 32. 2007, pp. 47.

Syed, Rahaman: «Los insectos polinizadores de la palma africana». *Revista Palmas*, 5(3), 1984, pp. 19–54.

Tardieu, Jean-Pierre: *El negro en la Real Audiencia de Quito. Ecuador S. XVI-XVIII*. Instituto Fránces de Estudios Andinos: Lima. 2006.

Tate, D. J. M: *The RGA history of the plantation industry in the Malay Peninsula*. Oxford University Press: New York. 1996.

Taussig, Michael: *Destrucción y resistencia campesina: El caso del litoral pacífico*. Punta de Lanza: Bogotá. 1979.

Tirado, Alvaro: *Introducción a la historia económica de Colombia*. Ancora Editores: Bogotá. 2000.

Tovar, Hermes: «La manumisión de esclavos en Colombia, 1809–1851: Aspectos sociales, económicos y políticos». *Revista Credencial Historia*, 59. 1994, pp. 34–41.

Uribe, Jaime: «Esclavos y señores en la sociedad colombiana del siglo XVIII». Retrieved 28.09.2014, from: http://www.revistas.unal.edu.co/index.php/achsc/article/view/29620

Usselmann, Pierre: «Geodinámica y ocupación humana del litoral pacífico en el sur de Colombia y en el Ecuador desde el Holoceno (últimos 10 000 años)». *Bulletin de l'Institut Français d'Études Andines*, 39(3), 2010, pp. 589–602.

Vargas, Ernesto: «Década del 60 y 70; la palma de aceite: de fincas a empresas». *Revista Palmas*, 23(3), 2002, pp. 86–92.

Vega, Gabriel: Comentarios al libro "La palma africana en Colombia. Apuntes y memorias". Revista Palmas. 23(4), 1998, pp. 15–18.

Walker, Brian "et al": " A handful of heuristics and some propositions for understanding resilience in social-ecological systems". *Ecology and Society*, 11(1), 13. Retrieved. 11.12, 2113, from http://www.ecologyandsociety.org/vol11/iss1/art13/

Wallerstein, Immanuel: *The modern world-system: Capitalist agriculture and the origins of the European world-economy in the sixteenth century*. Academic Press: New York. 2013.

Wallerstein, Immanuel: *La globalización o la era de la transición: Una visión a largo plazo del sistema mundo:* Paper presented at the Socialist Scholars Conference, New York City. 2010. Retrieved 16.04.2012, from: http://www.inisoc.org/waller64.htm

Wallerstein, Immanuel: "The rise and future demise of the world capitalism system. Concepts for comparative analysis". *Comparative Studies of Society and History*, 16(4), 1974, pp. 387–415.

Walsh, Catherine / León, Edizon / Restrepo, Eduardo: *Movimientos sociales afro y políticas de identidad en Colombia y Ecuador*. Universidad Andina Simón Bolivar: Quito. 2005.

Weber, Andreas: *Hybrid ambitions: science, governance, and empire in the career of Caspar G.C. Reinwardt (1773–1854)*. Leiden University: Leiden. 2012.

West, Robert: *Las tierras bajas del Pacífico colombiano*. Instituto Colombiano de Antropologia e historia: Bogotá. 2005.

West, Robert: *The Pacific lowlands of Colombia*. Lousiana State University Studies. Social Science Series Number Eight. Baton Rouge. 1956.

White, Nicolas, J: *British Business in Post-colonial Malaysia, 1957–1970*. Routledge: New York. 2004.

Whitten, Jr. Norman E.: *Black frontiersmen. A South American case*. Cambridge, MA: Schenkman Publishing Company. 1974.

Whitten, Jr. Norman E.: *Class, kinship, and power in an Ecuadorian town. The Negros of San Lorenzo*. Stanford University: California. 1965.

Whitten, Jr. Norman / Friedmann, Ninna: "La cultura Negra del litoral ecuatoriano y colombiano: Un modelo de adaptación étnica". *Revista del Instituto Colombiano de Antropología e Historia*, 17. 1974, pp. 81–115.

Williams, Eric: *Capitalism and slavery*. University of North Carolina Press: Chapel Hill. 1994.

Wilson, Charles: *The history of Unilever*. London: Cassel and Company LTD: London. 1954.

Wubs, Ben: *International business and national war interests: Unilever between Reich and Empire 1939–1945*. Routledge International Studies in Business History: London. 2008.

Zarembka, Paul: "Primitive accumulation in Marxism: Historical or transhistorical separation from means of production". *The Commoner*. Retrieved 30.04.2012, from www.commoner.org.uk/debzarembka01.pdf

Zuluaga, Francisco / Romero, Mario: *Sociedad y resistencia negra en Colombia y Ecuador*. Programa Editorial Universidad del Valle: Cali. 2007.

Reports ONG and official documents

Accenture for Humanity United (n.d.). *Exploitative labor practices in the global oil palm industry*. Accenture for Humanity United. Retrieved 20.05.2013, form http://humanityunited.org/pdfs/Modern_Slavery_in_the_Palm_Oil_Industry.pdf

Alcaldía de Tumaco: *Plan de Ordenamiento Territorial 2008–2019*. Tumaco: Alcaldía de Tumaco. 2008.

ANCUPA: «Ecuador País Palmicultor». *Revista Palma Ecuador.* 23. 2013. Retrieved 12.08.2014, from https://www.dropbox.com/s/q9ijs-4f0w9xg0u8/REVISTA-23-ESPECIAL-ok.pdf?oref=e

Corponariño: «Informe 2008». *Revista Hechos del Callejón*, 33. Marzo. 2008, pp. 15.

Comisión Intereclesiástica de Justicia y Paz: *La tramoya: Derechos humanos y la palma aceitera Curvaradó y Jiguamiandó*. 2005.

Corte Constitucional de Colombia, Sentencia T/025/2004. *Referencia: Protección de los derechos fundamentales de la población afrodescendiente víctima del desplazamiento forzado, en el marco del estado de cosas inconstitucional declarado en la sentencia T-025 de 2004*. Magistrado ponente Manuel José Cepeda.

Defensoría del Pueblo: *Violación de derechos humanos por siembra de palma africana en territorios colectivos de Jiguamiandó y Curvaradó*. Seguimiento de la Resolución Defensorial 39 del 2 de junio de 2005. Defensoria del Pueblo: Bogotá. 2006.

(2007, September). *Documento de las organizaciones participantes de la mesa redonda de palma sostenible*. Cali, Colombia. Retrieved 05.10.2014 from http://www.semillas.org.co/sitio.shtml?apc=e-b-20155545-20155545&x=20155568

FEDEPALMA: *Anuario Estadístico 2005–2009: La agroindustria de la Palma en Colombia y el mundo*. 2010.

FEDEPALMA: La agricultura de precisión: un cambio de paradigma en el manejo de las plantaciones. *Revista Palmas*, 29(2), 2008, pp. 5–6.

FEDEPALMA: «Productores defienden a la palma aceitera de falsas acusaciones». *El Palmicultor, Boletín Informativo de la Federación Nacional de Cultivadores de Palma de Aceite*, 410, 2006, pp. 22–26.

FEDEPALMA: *El cultivo de la palma de aceite en Colombia y el mundo*: Estadísticas 1989–1993. 1994.

FEDEPALMA: «Experiencia de Cordeagropaz en el modelo de alianzas estratégicas en el Municipio de Tumaco». Retrieved 16.05.2012, from http://antioquia.gov.co/PDF2/7%20EXPERIENCIA%20CORDEAGROPAZ-TUMACO.pdf

Friends of the Earth International: "The price of the palm. The fight for Indonesian forest and communities". *Friends of the Earth News Magazine*. 44(2), 2014. Retrieved 12.03.2015, from, http://www.foe.org/system/storage/877/f3/9/4933/Newsmagazine-Summer-2014.pdf

Insituto Investigaciones Ambientales del Pacífico: *Valoración integral del ecosistema de manglrar en el municipio de San Andrés de Tumaco (Nariño)*.Instituto de Investigaciones Ambientales del Pacífico: Quibo. 2012.

Malaysia oil palm Board: *Palm Oil Developments:* Kuala Lumpur. 1984.

MOE (2008). *Monografía político electoral Departamento de Nariño 1997–2007.*

Movimento Mundial por los Bosques: «*Condiciones de trabajo e impactos sobre la salud de los cultivos monoforestales a escala*». Retrieved 20.07.2013, from http://wrm.org.uy/fr/files/2013/01/Trabajo_salud.pdf

Oxfam: *Land and power. The growing scandal surrounding the new wave of investments in land.* 2013.

Oxfam: *Responsabiliad y sostenibilidad de la industria de la palma: ¿Son factibles los principios y criterios de la RSPO en Colombia? Dos Investigaciones.* INDEPAZ: Bogotá. 2010.

Presidencia de la República Colombia, Secretaría de Prensa: *Gobierno dispuesto a ser socio en proyectos de palma y biodiésel en Tumaco y Guapi, 2006. Retrieved 30.05.2012, from* http://oacp.presidencia.gov.co/snerss/detalleNota1.aspx?id=5381

Rainforest Action Network: *The human cost in conflict oil palm. 2015.* Retrieved 28.05.2013, from: *http://www.laborrights.org/publications/human-cost-conflict-palm-oil*

Tribunal Superior del Distrito Especial de Bogotá. Sala de Justicia y Paz. Magistrada Ponente Alexandra Valencia Molina (2014). Retrieved; 15.03.2015, from http://lasillavacia.com/historia/las-diez-cosas-que-prueba-la-sentencia-de-mancuso-49061

United Nation, Human Rights Council: *Report of the Special Representative of the Secretary General on the issue of human rights and transnational corporations and other business enterprises, John Ruggie.* 2013. (Human Rights Council, Seventeenth session, Agenda item 3 A/HRC/17/31). Retrieved 20.05.2014, from http://www.business-humanrights.org/media/documents/ruggie/ruggie-guiding-principles-21-mar-2013.pdf

United Nations Development Programme: "Human Development Report 2007/2008: The 21st Century Climate Challenge." Retrieved 14.08.2014, from http://hdr.undp.org/en/media/hdr_20072008_en_complete.pdf

United Nations Children's Fund: *Global inequality: Beyond the bottom billion: A rapid review of income distribution in 141 Countries.* New York. 2013.

WWF: *Peces dulceacuícolas del Choco Biográfico de Colombia.* Bogotá: WWF Colombia. Instituto de Investigaciones de Recursos Biológicos Alexander von Humboldt, Universidad del Tolima, Universidad Nacional de Acuicultura y Pesca, Pontificia Universidad Javeriana. 2012.

WWF: The oil palm industry in Malaysia: From seed to frying pan. WWF Malaysia. Retrieved 20.04.2013, from http://www.studymode.com/essays/The-Palm-Oil-Industry-In-Malaysia-776923.html

List of Interviews selected

1. San Lorenzo communitarian leader. Interview. November, 2012.
2. Afro-colombian women leader. Interview. November, 2012.
3. Former plantation worker San Lorenzo. Interview. December, 2014.
4. Farmer of Tumaco. Interview. December, 2014.
5. Farmer of Tumaco. Interview. January, 2013.
6. Farmer in Tumaco. Interview. January, 2013.

Index

A
Accumulation by dispossession 38, 39, 165
Africa 25, 38–40, 103, 141, 147, 150, 157, 179, 184, 187, 189, 191, 193
America
  Central and South America 11, 17, 56, 61, 63, 71–73, 166
ANCUPA 77, 95, 197
Andre Gunder Frank 19, 28, 35, 41, 171
Arturo Escobar 24, 52, 53, 52, 89, 95, 141, 148, 151, 155, 156, 159, 184

B
Banana 31, 72–75, 87, 95, 134, 137, 150, 154, 168
Biodiesel 58, 90, 192
Black-Africans 18, 19, 32, 33, 141, 143, 144, 148, 150–152, 158
Black-Afro communities 19, 25, 52, 55, 76, 78, 80, 82, 83, 85, 86, 88, 90, 134, 135, 140, 157, 159, 161, 176, 177
Botanic Gardens 60

C
Capital 36–41, 43, 44, 46, 50, 102, 103, 108, 109, 124, 137, 165, 170, 171, 181, 183–187, 189, 190, 192, 194–196
  capital accumulation 12, 16–18, 26, 28, 29, 32, 33, 33, 39, 41, 42, 44, 45, 47–51, 99, 100, 105, 115, 121–123, 128, 136, 153, 158, 167, 168, 175
  overaccumulation of capital, 38
  overaccumulation crisis 38, 165
Capitalism 37–41, 43, 46, 124, 165, 185, 186, 196
Capitalo-centrism 11, 21, 43, 46, 121
Chiquita Brand 72
  United Fruit 72, 73, 74, 168, 193
Colonial 15, 16, 19, 28, 29, 31, 32, 40, 59, 65–69, 77, 80, 98, 102, 117, 126, 127, 134, 140–147, 150, 154, 158, 162, 168, 172, 179, 187, 189, 190, 192, 194, 196
  colonialism 160
  neocolonial 167, 168
Commodity 18, 35, 49, 55, 59, 63, 69, 91, 102, 108–110, 113, 114, 117, 119, 121–123, 125, 172, 174, 183
Congo 56, 58, 61, 65, 67–69, 72, 74, 100, 187, 189
Cuadrilla 145–148, 151

D
David Harvey 37, 170, 187
Dispossession 16, 17, 21, 23, 25, 26, 33, 37–41, 43, 47, 50, 51, 68, 82, 86, 98, 105, 114, 117, 124, 159–161, 165, 167, 169–172
Dutch 60, 61

E
East India Company 61
Ecological system 124, 125, 134, 195

Economy 11, 16, 20, 23, 36, 45, 46, 48–50, 52, 66, 67, 79, 122, 135, 166
  market economy 32, 109, 140, 141, 157, 158
  political feminist economy 172
  world-economy 19, 28, 44, 49, 195
Ecuador Land Company Limited 29, 188
Elaeis guineensis 11, 56, 60, 71
Encomienda 34
Esmeraldas 24, 74, 143, 144, 179, 180, 184, 189–191, 193, 194
Ethnic, ethnicity 13, 34, 86, 154, 157, 158, 159, 166, 169, 174, 176, 177

F
Fauna 12, 132, 133, 151, 152, 157
FEDEPALMA 77, 78, 82, 91, 94, 95, 124, 197, 198
Feminist 39, 45, 46, 48, 166, 171, 172, 183, 184, 188
Fernand Braudel 28, 180
Friedrich Engels 26
Freedom 12, 16, 32, 49, 141–144, 147–152, 158, 176, 187

G
Gender 39, 146, 169, 171, 181, 185, 188
Giovanni Arrighi 27, 28
Guatemala 40, 41, 72, 186
Guerrillas 160

I
ICA 76, 77, 188, 192
Immanuel Wallerstein 28, 195, 196

Imperialism, imperialist 12, 99, 101–104, 179
Industrial Revolution 17, 29, 60, 64, 98, 167
Indigenous 20, 21, 28, 34, 40, 41, 52, 76, 78, 80, 83, 85, 86, 156, 174, 183
Indonesia 55, 61, 62, 68–72, 99, 179

J
J. K. Gibson-Graham 21, 44, 45, 121, 122, 135, 166, 186

K
Klaus Dörre 37

L
La Societé Anonyme des Huileries du Congo Belge (HCB) 65, 66
Labor, labor-power. 38–40, 46, 114, 115, 122–125, 154, 170, 171
  free labor 66
  forced labor 67, 71
  slaved labor 145, 148
  wage-labor 16, 23, 26, 33, 34, 36, 40, 44, 46–48, 50, 67, 98, 101, 111, 115, 122, 153, 166, 171, 172
Land 25, 38, 154, 179, 182, 198
  land enclosure 165, 167, 170
Land grabbing 11, 23, 25, 78, 165, 180, 191
LCA methodology 105, 106
Lever 65, 66, 67, 68, 69, 72
  Unilever 65, 73, 168, 184, 196
Life zones classification 130
Liza Grandia 40

M

Malaysia 55, 62, 68–72, 99, 179, 190, 196, 198, 199
Maroons 15, 148, 151, 162, 176
Marx 20, 26, 27, 35–39, 45, 114–116, 123, 137, 138, 161, 170, 175, 183, 185, 190
  Marxist-Marxism 16, 22, 27, 28, 32, 33, 35, 36, 39, 42, 44, 48, 122, 164
Massimo De Angelis 37
Metabolic rift 12, 13, 137, 139, 159, 161, 162, 185
Michel Foucault 43
Monocultural agricultural 59

N

Natural resources 25, 26, 29, 30, 33–35, 41, 45, 47, 52, 80, 83–85, 98, 109, 110, 112, 113, 115, 116, 118, 124, 128, 136, 147, 149, 151–153, 155, 160–162, 165, 166
Nature 12, 16–19, 25, 31, 32, 34, 36, 42, 43, 47, 49, 51, 52, 55, 59–61, 90, 95, 101, 102, 104, 105, 107, 109, 112, 115, 116, 118–128, 134–138, 153, 155, 159, 161, 162, 168, 170, 172, 174, 175, 177, 181, 185, 194

O

Oil palm 11, 23–25, 56–72, 74–82, 84, 86, 89–91, 94–97, 100, 106–108, 122, 124, 125, 134, 159, 161, 179, 181, 183, 186, 187, 189, 191–193, 197–199
oil palm crops 24, 65, 69
  palm oil 3, 12, 16–18, 23–25, 52, 55, 58, 59, 61–65, 67–80, 83–85, 87, 88, 90, 91, 94, 96, 99, 100, 103–114, 116–119, 121–123, 125, 128, 131, 133, 134, 136, 137, 149, 155, 159–162, 164–168, 172, 174, 198
  palm oil agribusiness 16, 17, 24, 25, 32, 53, 62, 73, 78, 97, 98, 100, 101, 107, 114, 123, 141, 155, 165, 167
  palm oil trade 17, 55, 63, 64, 65
  palma africana 24, 74, 76, 179, 182, 191–195, 197

P

Palenques 15
paramilitary 24, 78, 81, 87, 88 91
Primitive-primary accumulation 17, 26, 33, 35, 36–42, 170, 171, 183, 185

R

Racist-racism 80, 82
Relations of production 16, 17, 26, 27, 32, 35, 37, 40–44, 51, 113, 119, 122, 123, 128, 136, 158, 167, 171
  non-profit oriented relations of production 119, 129
Resilience 12, 19, 43, 51, 53, 124, 134, 137, 138, 141, 143, 151, 154, 159, 160, 162, 175, 176, 191, 195
  socio-ecological resilience 1, 3, 12, 53, 161, 121, 124, 161
Rosa Luxemburg 5, 26, 101, 189

S

San Lorenzo 4, 9, 11, 12, 15–17, 20, 23–25, 28–33, 35, 47, 50–52, 55, 56, 75, 78–91, 93–98, 100, 103–105, 107, 109, 111, 119, 121, 124–127, 129–134, 136, 137, 139, 141,

142, 145, 149, 154, 159, 161, 162, 165–169, 171, 179, 180, 189–191, 196, 199
Science-scientist 60, 62, 97, 104, 115, 117, 118, 154, 164, 168, 170
Second World War 31, 38, 68, 69
Silvia Federici 39
Slave trade 63, 64, 137, 142, 150, 167, 176, 193
Smallholder 70, 71, 95, 96
Social metabolism 12, 48, 124, 137, 138, 153, 157, 160–162, 190
South Asia 11, 56, 63, 68, 70, 71, 74
Surplus-value 18, 26, 27, 33, 37, 40, 44, 101, 103, 113–116, 122, 123, 147, 160, 171

T

Taussig 29, 135, 141, 195
Tumaco 9, 11, 12, 15–17, 20, 23–25, 28–33, 35, 47, 50–52, 55, 56, 75, 78–84, 87–91, 94–98, 100, 103–105, 107, 109, 111, 112, 119, 121, 124–127, 129–137, 139, 141, 142, 145, 149, 154, 159–162, 165–169, 171, 179, 181, 182, 184, 187–189, 197–199

U

Unequal exchange 17, 18, 34, 49, 101, 104, 105, 114, 119, 126, 136, 153, 160, 168, 175, 190
USA (United States of America) 72, 94, 98, 102, 103, 117, 166, 169, 186

W

West Africa 11, 17, 25, 56, 60, 63, 64, 70, 72, 99, 100, 104, 152, 189
White-mestizo 80, 92, 150, 152, 154, 155
Worker 16, 27, 46, 47, 68, 91–93, 110, 112–114, 124, 166, 167, 172, 199
World Bank 155
World-system 28, 40, 88, 100–102, 104, 105, 115, 124, 126, 136, 152, 153, 166, 169, 183, 195

**Transpekte: Transdisziplinäre Perspektiven der Sozial- und Kulturwissenschaften /
Transpects: Transdisciplinary Perspectives of the Social Sciences and Humanities**

Herausgegeben von Johannes Angermüller, Anke Bartels, Dietmar Fricke,
Raj Kollmorgen, Jörg Meyer und Dirk Wiemann

Band 1 Dirk Wiemann / Agata Stopi•ska / Anke Bartels / Johannes Angermüller (eds.): Discourses of Violence – Violence of Discourses. Critical Interventions, Transgressive Readings, and Postnational Negotiations. 2005.

Band 2 Michael Schultze / Jörg Meyer / Britta Krause / Dietmar Fricke (Hrsg.): Diskurse der Gewalt – Gewalt der Diskurse. 2005.

Band 3 Anke Bartels / Michael Schultze / Agata Stopi•ska (eds.): Re/defining the Matrix. Reflections on Time – Space – Agency. 2006.

Band 4 Britta Krause / Tania Meyer / Nina Pippart / Dietmar Fricke (Hrsg.): Chronotopographien. Agency in ZeitRäumen. 2006.

Band 5 Agata Stopi•ska / Anke Bartels / Raj Kollmorgen (eds.): Revolutions. Reframed – Revisited – Revised. 2007.

Band 6 Anke Bartels / Reinhold Wandel / Dirk Wiemann (eds.): Only Connect: Texts – Places – Politics. Festschrift for Bernd-Peter Lange. 2008.

Band 7 Dirk Wiemann / Lars Eckstein (eds.): The Politics of Passion. Reframing Affect and Emotion in Global Modernity. 2013.

Band 8 Verena Rodatus: Postkoloniale Positionen? Die Biennale DAK'ART im Kontext des internationalen Kunstbetriebs. 2015.

Band 9 Edna Yiced Martinez: Capitalist Accumulation and Socio-Ecological Resilience. Black People in Border Areas of Colombia and Ecuador and the Palm Oil Industry. 2018.

www.peterlang.com

www.ingramcontent.com/pod-product-compliance
Ingram Content Group UK Ltd.
Pitfield, Milton Keynes, MK11 3LW, UK
UKHW021834210426
5322IPUK00018B/259